Assembly
with Robots

Assembly with Robots

Tony Owen

PRENTICE-HALL, INC.
Englewood Cliffs, New Jersey 07632

First published in 1985 by Kogan Page Ltd
120 Pentonville Road, London N1 9JN

This edition published 1985 by
Prentice-Hall, Inc., Englewood Cliffs,
New Jersey 07632
ISBN (Prentice-Hall edition): 0-13-049578-6

Printed in Great Britain

Contents

Preface

In the western world, economic logic (and need) has replaced the indentured craftsman by computer controlled machining centres within manufacturing industries.

The same rationale is the incentive behind the development of robots that are technically capable of performing assembly tasks, and the inevitable, albeit slow, adoption of these robots by the manufacturing industries.

This book is based upon the author's knowledge and first hand experience of the manufacturing industries of North America and the UK in general, and the UK's robotics industry in particular. The general and specific implications of performing an assembly task robotically are discussed, the majority of which are not specific to any one sector of the manufacturing industry, nor to any particular size of product being manufactured. This book should be of interest to those who are interested in or involved with the use of robots for assembly. The 'veils of mystic' and misinformation on robots and the assembly process are subsequently removed.

The author wishes to acknowledge the photographs and information provided by the following people and companies: ASEA Ltd; ASEA Robotics (Västerås); Battelle Research Laboratories; Geoffrey Champion; Cranfield Robotics and Automation Group, Cranfield Institute of Technology; Concentric Production Research Ltd; Danichi Sykes Robotics Ltd; Evershed Robotics Ltd; Fairey Automation Ltd; GEC Research Laboratories (Marconi Research Centre); George Kuikka Ltd; Dr John Hill, University of Hull; IBM UK Ltd; John Brown Automation Ltd; Martonair Ltd; Metzeler & Lord Gimetall Ltd; PA Technology; Prutec Ltd; Dr Alan Redford, University of Salford; Rediffusion Robot Systems Ltd; Sale Tilney Technology PLC; 600 Fanuc Robots Ltd; Dr Ken Swift, University of Hull; Swissap; Telelift GmbH & Co; Unimation (Europe) Ltd; VS Remek Ltd; Zentel Ltd.

CHAPTER 1
Introduction

Assembly with robots represents the leading edge of technological development in today's manufacturing industry. If robotics
and assembly are considered separately, then assembly (as an
automated process) has been in existence for almost 100 years,
and robotics for about a quarter of that time. The use of robots
to perform assembly tasks has been the subject of much
research since the mid 1970s, but only recently has it been shown
to be both technically and economically viable within the general manufacturing environment. As with advances in other
technologies, the real world of industry lags far behind the
frontiers of research into robotized assembly, because the 'new
discoveries' need to be both proved and adapted to industry.

An assembly task occupies 53% of time and 22% of labour
costs involved in manufacturing a given product. Much of
today's assembly work is performed manually, or by automated machines. Manual assembly is used for low demand
products and/or to perform tasks that are beyond the present
capability of technology. Automated machines are used to
satisfy large demand levels, and are restricted to the assembly
of single groups of products.

The term 'assembly' is used here to mean the joining together
of a number of discrete items to form a composite item (eg
welding, screwing, glueing or fitting parts together, all of
which are used in the manufacture of many of today's consumer
products). However, assembly *could* also mean palletizing, in
that a number of items are loaded onto a pallet in a particular
arrangement until the load is completely 'assembled'.

With such a wide range of processes, the use of robots for
arc welding, spot welding and pallet loading has been omitted
from this book. The reasons are threefold:

1. The accuracy and repeatability of the robots (generally)

11

used for these tasks is in excess of the arbitrary 0.05mm limit adopted with this text.

2. The robots and control systems used for arc welding have a number of very specific requirements not applicable to other assembly tasks.

3. The major requirement for robots used for spot welding and palletizing is that of brute strength, which is not a significant criterion within the generally understood meaning of robotic assembly.

As an example of the problems of attempting to define and limit what constitutes an assembly robot, consider VS Engineering's robotic solution to the assembly of a windscreen into an automobile on the Montego line at Cowley, UK. The glass for the windscreen has to be positioned within a tolerance of 1.00mm. The total weight of gripper and the windscreen demands that a Unimate 4000, which has a repeatability of $+/-2.03$mm, is used. Technically, the solution can be considered to be a 'robot carried by a robot', with the Unimate being used to position roughly the gripper and glass (Figure 1.1). With the Unimate static, the intelligent gripper containing four linescan cameras, associated lighting and stepper motor systems (in addition to the vacuum cups) moves the glass vertically, horizontally and/or rotationally to match the window aperture. The gripper then inserts the windscreen into the bodyshell.

Robots have gradually evolved from primitive manipulators with virtually zero intelligence, to devices that have high levels of 'pseudo intelligence' that can communicate with other machines and react to changing work environments. The development and widespread availability of microcomputers have permitted the development of very sophisticated control and sensing systems compatible with all but the most simple of robotic devices.

Before installation, it is important to ensure that the assembly tasks are compatible with the robot's capabilities, otherwise failure both as a technical project and a financial investment will result. In addition, the product being assembled has to have been designed for automation. An automated process requires that all activities are provided with the necessary materials and that the activity is inspected to ensure that it has been performed satisfactorily at both physical and functional levels.

Although many assembly tasks can be performed by robot,

Figure 1.1 *A Unimation Unimate 4000 robot is used to assemble a windscreen into an automobile's bodyshell. The use of an intelligent gripper more than doubles the operational precision of the robot.*
(courtesy of VS Engineering Ltd)

there are others that need the skills and intelligence that only man can (at present) provide. The mixing of robots and humans on an assembly line should be considered as only an extension of the interaction of men and machines in the traditional manufacturing industry. Safety is the one fundamental work difference between people and robots, which must be acknowledged and catered for in every installation.

Even though robots may be thought of as 'just' machines, they do have special features that distinguish them from other machines and make them potentially more dangerous to work with. For instance, the robot's work envelope is usually *outside* its base, making it difficult to 'visualize' the safety zone; robots are occasionally seen as 'motionless', yet without prior warning they may start moving within a three dimensional space, thus being potentially dangerous to a casual observer.

Task performance can be measured by technical achievement and economic viability. In the real world, where competitors and fractions of a penny mean the difference between profit, survival or bankruptcy, there are few circumstances where technical success takes precedence over economic sense. The economics of using robots to perform an assembly task are complex. Essentially, they involve both quantitative and qualitative values for a number of tangible and intangible benefits that *should* result from the adoption of the robotic solution.

There are many elements that make up and complicate an assembly task. Their importance to the success or failure of a project is discussed in the following chapters and appropriate observations are made.

CHAPTER 2

Why use robots?

There are a number of 'approved' definitions for robots. However, in general, a robot is a reprogrammable manipulator capable of moving various objects through variable programmed motions.

A refinement that takes into account the control sophistication of top line robots would be that 'A robot is a reprogrammable manipulator fitted with integral sensory perception that allows it to detect changes in the work environment or work condition and, by its own decision-making faculty, proceeds accordingly'.

Robot versus hard automation

It is generally considered that robots, when compared to humans, yield more consistent quality, more predictable output, and are more reliable. However, when they are compared with automated machines, then their flexibility becomes quickly apparent compared with the rigidness of hard automation. A robot is essentially an arm fixed to a base upon which it can move, the preciseness and range of motion being dependent upon the sophistication of the robot's control system. In contrast, an automated machine's motions are fixed and generally have no redundant degrees of freedom to allow it to process products outside its narrow range.

The robot's flexibility is illustrated in Figure 2.1, which shows how a 'standard' robot can be reconfigured (through specialist software and hardware accessories) to perform a vast range of tasks. It therefore makes economic sense to purchase a robot whose 'optional' features are relatively low priced when compared to the robot's base price, rather than a different, specialized machine for each task. However, the robot, being a generalist device, has many redundant links and features that result in a slower process time than a machine designed specifically for the task would take. For example, the task might require the use of only four or five of the robot's six axes of motion. Another example is where a servoed jointed arm robot is used to perform a simple straight line insertion task, instead of a single axis device.

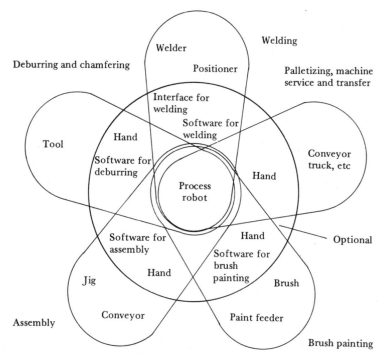

Figure 2.1 *The use of appropriate software and hardware accessories allows a 'standard' robot to be customized to a specific task. The robot's flexibility is limited only by its size, payload, repeatability and degrees of freedom.*
(from *Hitachi Review*, Vol.30, No.4, p.209, 1981)

Nowadays a typical product life cycle is generally considered to be two or three years, instead of the historical five to ten years. Consequently, many manufacturing lines will be scrapped or reconfigured after three years. A flexible system, incorporating robots, allows a higher percentage of items to be salvaged, unlike hard automation which is rarely salvageable and the return on investment (ie cost of developing and installing the system) has to be recovered before the system becomes obsolete. In addition, because the system is flexible within its geometric, power and functional constraints, its life expectancy is longer than that of hard automation.

Assembly-related tasks take many forms (eg the placement of components, the laying of a sealing ring, or the installation and fixing of bolts) and through the use of specialist grippers and/or interchangeable tools, a single standard robot *could* be used to perform any task. In contrast, it is usually more expensive to develop a dedicated automation system, and it is doubtful if it would be any more successful than the robot. Further, because the robot is an 'off-the-shelf' item, its inherent reliability is usually better than that of the custom-designed equipment, even though the specially designed attachments will reduce overall system efficiency.

The task a robot performs is described more in software than hardware; this allows a wide range of similar products or tasks to be processed with the minimum of reconfigurations to its workstation. For instance, if a number of products needed sealant applied in different patterns (ie missing holes and other surface features), then one robot fitted with a sealant dispenser could process a number of different items. To ensure success, however, it is important that the items are identified by a 'computer schedule' before commencement of the process, or be identifiable through bar codes or other sensory information that enables the robot to select the correct dispensing program. Additionally, each type of item being processed should be presented in a constant attitude and orientation. The use of location and clamping devices, that push or pull the items against fixed datum edges, readily achieves this.

A robot is generally a single arm functioning within a three dimensional work envelope normally uncluttered by the structure of the robot. The task is therefore performed in a situation that allows maximum access. Not so with hard automation, which invariably performs tasks within the structure of the machine, where easy access and observation of the task are not always possible.

Robot availability, purchase and viability
Nowadays, with the glut of suppliers, robots can be purchased 'off the shelf'. Therefore, if an existing system needs to be expanded, the robot element, along with the grippers and tools, can be quickly manufactured or purchased. It is acknowledged that the interfacing and programming might take time

to debug, but overall an addition to, or a duplication of, an existing system can be achieved in a shorter time than could the extension or duplication of a hard automation system.

The availability of robots also has a positive advantage in maintenance terms, because replacement units are available almost *ex stock*, and because the robots are designed as modular units, individual modules can be obtained at short notice.

At present, the robot market is saturated with suppliers trying to market their products to a small and limited audience. This has two effects:

1. The unit price of robots is dropping, or can be haggled down from list price.
2. The robot market is concerned about the viability of the companies from whom they purchase their robots.

The concern over robot viability is a real one, with some 32 companies (at the moment) in the UK marketing around 91 types of true robot. In addition, there are 20 suppliers of pick and place units and four of educational robots. Of these, less than six are indigenous robot manufacturers, the remainder having franchises, or being offshore robot manufacturers. The situation is similar in the USA, with over 100 companies involved in the robot market. In Japan, however, the number is about 140, though their products are not always fully servoed devices.

The problem with giving the number of vendors of robots, or market size, is that the information is quickly out of date because of the dynamics of this particular market. However, the information is still valid and essential for any potential user of robots as it indicates the faith companies have in their products, and any possible risks to the purchaser, ie the risk of the vendor going out of business.

For the potential user of assembly robots, there are (at the moment) some 17 primary suppliers in both the UK and USA (see Appendix). The number of robot suppliers grew dramatically during 1980/81 and has been fairly constant since, even though the companies in the vendor group have changed significantly, with some leaving or entering freely according to market changes or identification of apparent niches. Other 'names' have changed through takeovers, or bankruptcy.

It is the considered opinion of many within the robot industry that there will soon be a big 'blood-letting' and the market will settle down to approximately 10 suppliers, of whom the present larger vendors will form the majority, to supply a bigger and more confident market.

Figure 2.2 shows the robot population of the world in 1983,

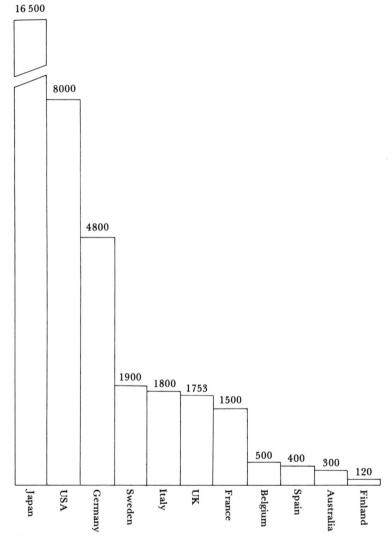

Figure 2.2 *The world population of 'true robots' as of December 1983. The census requested that participants adhere to devices that fitted the definition, since there are many devices used within industry that approximate some, but not all, of the attributes of true robots.* (from British Robot Association 1984)

whilst Figure 2.3 shows the increase of robot population within the UK and West Germany over the last decade; the number of robots used in assembly-related tasks in these two countries is shown in Figure 2.4. The increasing awareness and adoption of robots for assembly are further shown in Figure 2.5. (Data for the USA and Japan is either not available or is unreliable.)

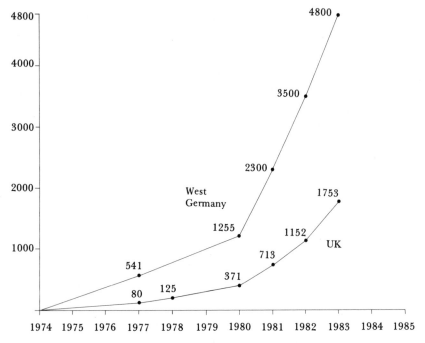

Figure 2.3 *Growth of the robot population in West Germany and the UK since 1974. Prior to 1980, interest was limited and parochial, hence the scarcity of data.*
(from British Robot Association 1984)

Assessing the robot market

Every census is based upon the vendors' interpretation of the questionnaire, the truth of the answers and the number who bother to answer, complete and return it. To assess a particular sector of the robot market, it is important that the data used is valid, correct and up-to-date. Table 2.1 shows the result of two surveys on the same market during the same time frame, and the potential risk of error can be seen quite clearly. It can also be seen that whilst the total quantity is the same within 2%, the number for 'assembly tasks' varies by 1000%. It follows there-

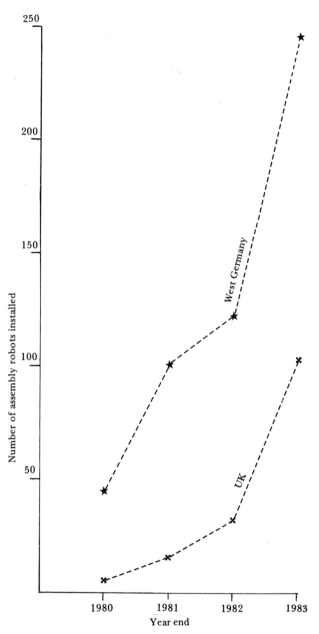

Figure 2.4 *Growth in the use of robots for assembly tasks in both West Germany and the UK since 1980. As with Figure 2.3, the data is correct as of December of each year. In pure numbers, West Germany has almost 2½ times the quantity of installed assembly robots of the UK.*

21

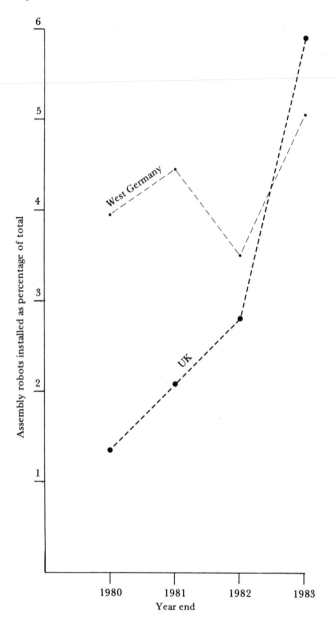

Figure 2.5 *The data used in Figure 2.4 is converted into a percentage of the total number of robots in West Germany and the UK. The difference between the 1982 and 1983 figures is about the same even though the order was reversed. The exact reason for the 'dip' in the West German figures is unknown, though Table 2.1 possibly holds a clue.*

Task	Diawa Securities	RIA
Foundry	615	840
Spot welding	1,435	1,500
Loading	820	850
Assembly	410	40
Painting	615	640
Others	205	400

Table 2.1 *Estimated robot population of the United States by Task in 1981* (From Economist Intelligence Unit, London, Report Number 135, *Chips in Industry*, 1982.)

fore that classification is based upon a subjective interpretation of the tasks involved. Under these arbitrary rules, would 'fettling' come under 'foundry' or 'machining'?; at what degree of accuracy does palletizing/packing become assembly?; and is 'enamelling' the same as 'painting' or should it be placed under 'other'?

It is not enough to know the reasons why robots should be used when deciding which particular robot is suitable for which assembly task. To help in arriving at the correct decision, the various features that describe a robot in terms of its performance and suitability are discussed in Chapter 3.

Which configuration?

There are presently six recognizable configurations of robots: polar, arm and elbow, cylindrical, cartesian, gantry and SCARA (Figures 3.1-3.6). Kondoleon (1976) examined assembly tasks in terms of direction of work activity and found that a single vertical motion accounted for the majority of tasks; in other words, assembly consists of single linear motions that move components vertically downwards from a rest position above an item, so that the two objects are in physical contact.

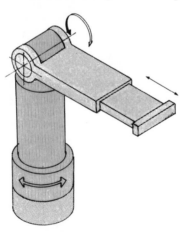

Figure 3.1 *Polar configuration robot. The body rotates vertically, the arm raises or lowers in the horizontal plane and extends radially.*

Other motions associated with assembly tasks are the pick and place activities that transport the individual items from presentation devices to the preassembly position mentioned earlier. In general, nonlinear and nonvertical motions are few and far between, and their use within assembly tasks is more a function of the geometry of the robot rather than the requirement of the assembly task.

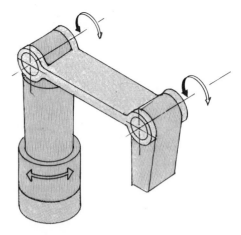

Figure 3.2 *Arm and elbow configuration robot.*
The body rotates vertically. The 'arm' has two joints,
which have independent movement in the horizontal plane.

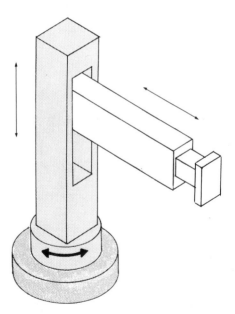

Figure 3.3 *Cylindrical configuration robot. The body rotates vertically,*
whilst the arm has both vertical and horizontal motion.

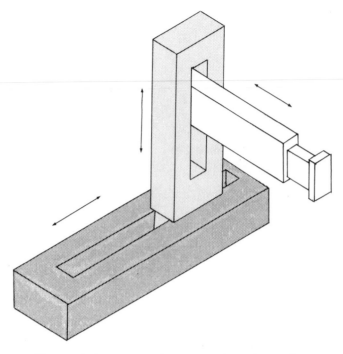

Figure 3.4 *Cartesian configuration robot. This unit has three perpendicular linear motions.*

In essence then, the ideal robot for an assembly task is one that has vertical motion which can service a given horizontal surface. Cartesian and gantry robots match this criteria, but the SCARA robot (see Figure 3.7) is also suitable as it offers vertical motion from any point within a work plane, even though its horizontal work plane is a series of curves.

Cylindrical configuration robots (Figures 3.8 and 3.9) also offer a horizontal work plane from which vertical assembly motions can be made. Its advantage is that direct horizontal moves can be made through simple arm extension, which could be essential in a crowded workstation. The cylindrical and cartesian configurations are similar and their actions are simple, yet they are used less frequently than the arm and elbow, or SCARA.

If the cartesian and gantry robots offer the best match to the stated attributes of an assembly task, then why is the PUMA robot (Figure 3.10), an arm and elbow configuration, so popular as an assembly robot? The reasons are manifold, the two most important being:

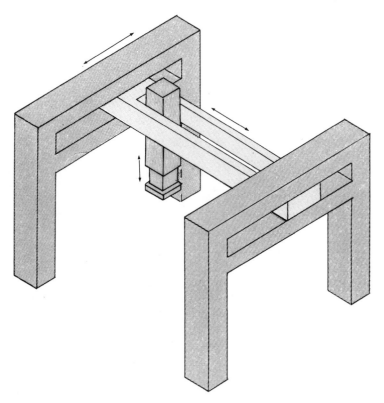

Figure 3.5 *Gantry configuration robot. It has identical motions to the cartesian configuration, but includes an additional support. This means that the arm is not cantilevered, so the unit has greater rigidity.*

1. It was designed to duplicate the motions of the human assembler so that it could (theoretically) be used for assembly tasks without much redesign of the workstation.
2. Its work envelope is more comprehensive than the cartesian, gantry or SCARA in that it has 'front', 'back' and 'sides'. This means an assembly task can be performed at the 'front', whilst components are stored and collected from the back. This is analogous to a person in a workstation with storage containers and tools placed to the side and back.

Both the PUMA and SCARA have revolute joints to enable a multi-axis movement which is often faster than the three-axis prismatic motions of the cartesian and gantry units. However, the control system for the PUMA is more complex than the

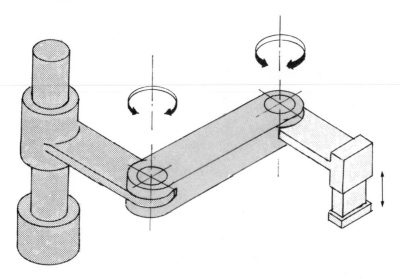

Figure 3.6 *SCARA configuration robot. A three linked arm with vertical servoed rotation. The gripper has a limited and usually nonservoed simple vertical action.*

cartesian and gantry robots, mainly because the moving of the 'gripper' in a vertically downwards motion requires high speed computation to transform all the revolute joint motions into a single precise, smooth vertical motion. Cartesian robots (eg DEA's PRAGMA, Figure 3.11; or VS Remek's PAM, Figure 3.12) are structurally stiffer than the revolute joint robots, though they are not as stiff as the gantry configuration (eg IBM 7565, Figure 3.13).

The available polar configuration robots are 'heavy duty' manipulators and are not generally used for assembly tasks. The exception shown in Figure 1.1 was used on the basis of its payload capacity.

Ultimately, the choice of robot configuration is a mixture of objective technical reasons (eg repeatability, payload and work envelope) and subjective reasons such as price, the number of that manufacturer's robots that are in service, and the rapport that is developed between the vendor and buyer.

Specialized robots
There are a number of specialized robots whose construction does not fit into any of the six configurations. For instance, the Microbro MR-01 Souris, which as shown in Figure 3.14 has a work envelope in the form of a tilted spherical zone. This

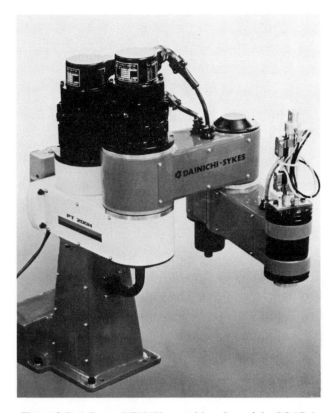

Figure 3.7 *A Daros PT200H assembly robot of the SCARA configuration. Horizontal motion is servo controlled, whilst the standard vertical motion is pneumatic. This unit has a payload of 2kg and a maximum horizontal reach of 600mm. This robot is also available as the PT200V, which moves its arms in a vertical plane.*
(courtesy of Dainichi Sykes Robotics Ltd)

electric robot is, in simple terms, a very sophisticated wrist/ gripper unit supported by a manually arranged arm and fixed in the required position. The unit is extremely precise (0.02mm), being specially designed for the 'microtechnical' tasks, and is very fast.

PA Technology are presently marketing the YES-MAN robot (Figure 3.15), a double-armed device whose arms function either independently or interdependently. One arm can be used as a 'jig', while the other performs tasks on the component held by the jig. This robot was developed and funded by Prutec and has yet to find a commercial manufacturer. The reasons no doubt stem from market uncertainty.

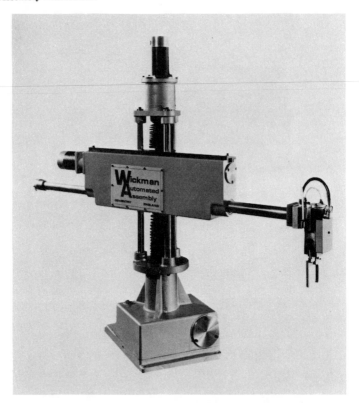

Figure 3.8 *The Wickman W500 robot has a cylindrical configuration, and a payload of 10kg and a maximum reach of 1000mm. With the exception of the wrist, all axes are servo controlled.*
(courtesy of John Brown Automation Ltd)

NONSERVOED PICK AND PLACE UNITS

A number of devices, not usually classed as robots but fully capable of performing many assembly tasks, are the nonservoed pick and place units. These units are usually in one of the three configurations (ie cartesian, gantry or cylindrical) and can be supplied as modular parts that can be configured into a desired arrangement (see Figure 3.16).

The axes of these nonservoed units have adjustable mechanical stops that set and limit motions. Generally, each axis has two stops to define the limits of the set stroke of an axis for a particular task. Certain devices are supplied with multiple stops for each axis. For instance, extreme stops are fixed and the intermediate stops are blocks, each containing a solenoid

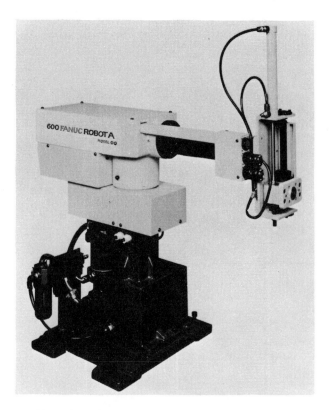

Figure 3.9 *The smallest of a range of three cylindrical assembly robots, the Fanuc Model A00 has a payload of 10kg and a maximum reach of 582mm.*
(courtesy of 600 Fanuc Robotics Ltd)

operated pin. These pins are then normally retracted and, only when the program requires that they form the limit for the 'next' stroke, are they activated to protrude from their housings and act as the end stop. The provision of multiple stops on each axis allows nonservoed devices to approach the flexibility of servoed robots. The only difference is that the servoed robots have an infinite number of stroke settings within the extreme limits of each axis, whereas the available strokes for any one task on nonservo devices are restricted to the number of stops installed (ie one axis with two (extreme) stops has two strokes, one in each direction as it moves from one extreme position to the other). Figures 3.17 and 3.18 show two examples of these devices.

Figure 3.10 *The 'original' assembly robot, the PUMA 260, has an arm and elbow configuration. Its payload is 0.9kg and its maximum reach is 406mm. This is the smallest of a range of five models.*
(courtesy of Unimation (Europe) Ltd)

Another disadvantage with the nonservoed units is that the stops have to be manually fixed and adjusted. This can be time-consuming since both the stroke length and terminal positions need to be positioned accurately. Nonetheless these devices have three distinct advantages over servoed robots:

1. Repeatability is much better than a 'true' robot, typically $+/-0.02$mm versus $+/-0.05$mm for the majority of commercially available assembly robots.
2. They are faster in that they are pneumatically operated and 'decelerate' by smashing against the end stops, which are cushioned to absorb the energy of impact so no damage is done to the system.

Figure 3.11 *DEA's A3000 PRAGMA is a cartesian configuration robot, designed for multirobot applications. Four units can be operated from one controller. In this experimental setup, two PRAGMAs are assembling compressor valves.*
(courtesy of Fairey Automation Ltd)

3. Cost is between one-tenth and one-half of that of a servoed assembly robot.

Robot capability

Irrespective of configuration, the robot chosen for an assembly task must be capable of achieving the technical parameters of payload, repeatability, time of action and reach for a particular task, as well as being easily programmable. Unfortunately, whilst the technical specifications of robots usually include this information, they do not state the conditions under which these values were obtained. For example, the velocity and repeatability of an arm and elbow configuration must depend upon the arrangement of the arm, since the velo-

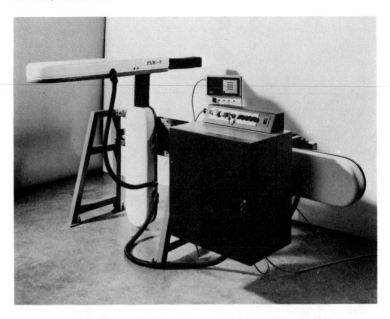

Figure 3.12 *A UK designed and manufactured cartesian robot, the VS Remek PAM 2. Its microcomputer controlled drive system uses pneumatics for high speed motion and stepper motors for fine positioning.*
(courtesy of VS Remek Ltd)

city (if given in m/secs, as is usual) varies with the effective arm radius. Similarly, the repeatability for a revolute system is an angular function and the usually stated linear values are not applicable unless the testing configuration and correlation data is known.

As the payload carried by a robot varies, so the time taken to reach a given velocity must also vary. Further, the stated slew velocity is not always reached if the axis motions are small. Calculation of the cycle time expected for a given task is outlined in Chapter 4.

Robots can be programmed in a number of ways, which can negate or magnify the problems of vague and incomplete robot specifications (see below).

Which programming method?

Robots are programmed either online at the actual workstation, or offline at some computer terminal.

Figure 3.13 *IBM's 7565 gantry configuration robot.*
A rigidly designed hydraulic robot, though less precise than any
of the other assembly robots listed in this book.
(courtesy of IBM United Kingdom Ltd)

ONLINE PROGRAMMING
This type of programming is performed by one of three methods: lead through, drive through or coordinate entry.

Lead through programming
This method is the oldest form of programming, which requires the programmer to move the robot through the desired pattern manually. The robot records the coordinates of the end-points of each discrete motion, or samples the pattern in real-time and records coordinates at fixed time intervals.

Lead through can be used for actions where some 'feel' is needed to achieve a correct position. For instance, if a robot is to fit two parts together, then the exact horizontal positions that allow a 'free vertical motion' must be sensed by the programmer.

Drive through programming
The programmer in this method uses the teach pendant to drive the axes independently, or simultaneously, through the

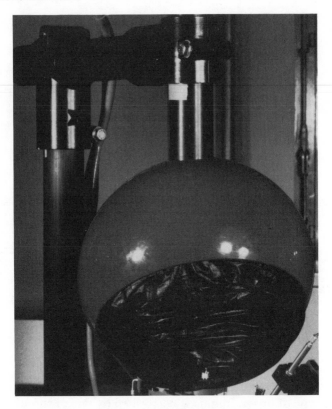

Figure 3.14 *Microbro's MR-01 (Souris) robot has a virtual hemispherical work envelope. Its payload is 100g and its maximum programmable reach is 190mm. This unit is cantilever mounted from a rigid post that supports it over the work zone. The software reprogrammable motions duplicate those of the human wrist and hand in that they can pivot about a 'central point', extend and contract, and grasp.*
(courtesy of Concentric Production Research Ltd)

pattern. Again the coordinates can be recorded as end-points, or sampled in time.

This method may also be used to program the mass of the pattern, without requiring the programmer to 'haul' the robot around.

Coordinate entry
This method uses a pendant to enter coordinates for each action, without moving the robot.

Coordinate entry is used best as an editing feature to adjust some of the coordinates, put in conditional moves, program

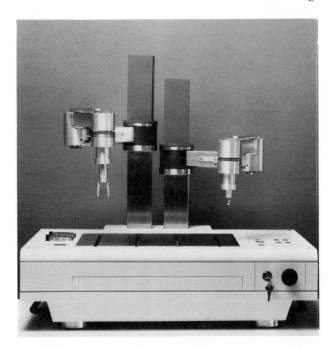

Figure 3.15 *The YES-MAN robot, whose design and manufacture by PA Technology was funded by Prutec Ltd, has unfortunately not yet found a commercial vendor. The concept of two arms that can function independently or interdependently is, as yet, unique in that it duplicates the effective motions of the human assembler.*
(courtesy of Prutec Ltd and PA Technology)

the interlocks and to operate the grippers as the robot is being programmed, using the other modes.

OFFLINE PROGRAMMING

Offline programming is performed at a computer terminal away from the workstation. This means that the robot can be used to perform value added work, whilst other programs are being developed and simulated. This mode of robot programming is identical to computer programming in that flow diagrams are developed and transferred into programs that include all interlocks, instructions to grippers and feeders, and test routines.

After simulation, to ensure that the program does not collide with any (database) structures, or does not conflict with the robot's physical and control parameters, the program can be downloaded into the robot for running. The correlation between successful simulation and successful real world opera-

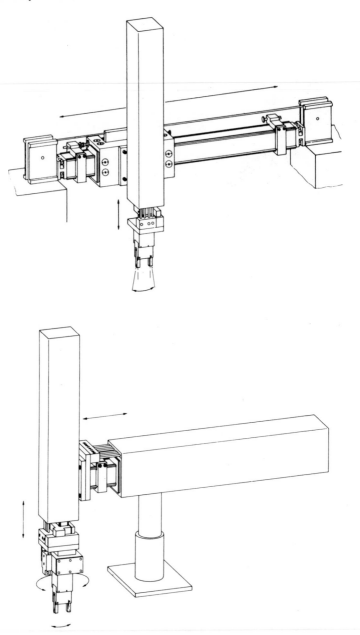

Figure 3.16 *Nonservoed modules can be fitted together in many combinations to provide a simple pick and place device.* (courtesy of Martonair Ltd, Twickenham)

tion is a function of the correctness and completeness of the database used.

Important elements in robot specifications
The three most important elements in a robot's specification are payload, repeatability and accuracy, since they define what can be carried and how precisely it may be manoeuvred.

PAYLOAD
Payload is the weight that can be carried at the end of the robot's arm. Traditionally, robot specifications have not included the gripper. Consequently, payload must include the gripper and wrist-gripper interface, as well as the maximum weight component to be manipulated.

Several robots, specifically designed for assembly tasks, are supplied with a standard gripper. In this case, the stated payload is of course the maximum weight that can be carried by that robot-gripper arrangement.

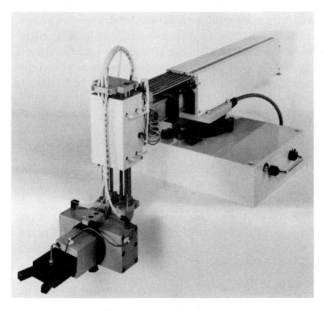

Figure 3.17 *A three axis nonservoed device*
assembled from discrete modules.
(courtesy of Martonair Ltd, Twickenham)

REPEATABILITY AND ACCURACY

Repeatability and accuracy are two terms that are often used incorrectly in specifications. Repeatability is the range within which the robot will terminate a motion and accuracy is the error between the targeted position and the mean repeatability (Figure 3.19).

Accuracy is a complicated measurement in that it is affected by the following:

1. Squareness and parallelism of the robot structure.
2. The geometric relationship between robot and the worktable upon which the assembly task will be performed.
3. The stiffness of the robot.
4. The variation of weights that the robot is transferring.

Except for offline programming, accuracy is unimportant. The reason is that the robot is programmed at the workplace and all errors due to out-of-squareness, non-parallelism, variation in component weight, deflection of robot structure, etc, are neutralized by online programming.

Offline programming assumes that all information about the robot and its workplace is known and correct. Consequently, any errors in the database will cause inaccuracies in robot performance and could cause damage. There are two alternatives: the first is to

1. Use offline programming purely as an economic means of generating and simulating programs without tying up a robot. However, when those programs are first downloaded to the robot, it should be 'single stepped' through the program and any variances corrected by online coordinate entry.
2. Ensure that the information in the database is correct for each and every robot and workplace. This is a long-term policy, developed from compiling the variances encountered through using the first approach.

TASK TIMES

For a given task, robots are slower than both hard automation and humans. Over a shift, because of the variances in human performance, the robot is generally more productive (quicker) than humans, but is still slower than hard automation. The time taken for a robot to perform a task depends upon the

Figure 3.18 *The MHU Senior represents the leading edge of nonservoed manipulators. Although it is more expensive than some assembly robots, its motions and control system enable it to match the general capability of many 'true robots'.*
(courtesy of George Kuikka Ltd, Watford)

length of movement, the number of axes involved and the robot configuration.

Although robots have stated maximum velocities, they have to be accelerated to these velocities. If the motions are short, then the achieved velocity is less than the maximum. Also, because each move requires that the robot be accelerated and decelerated, the average velocity for that motion is much lower than the maximum listed. If a move involves more than one axis, then again the combined velocity is different.

Robots with revolute joints often move in arcs between one point and another. Although the listed arm velocities for these robots are higher than those given for robots with prismatic motions, it should be remembered that an arc drawn between

41

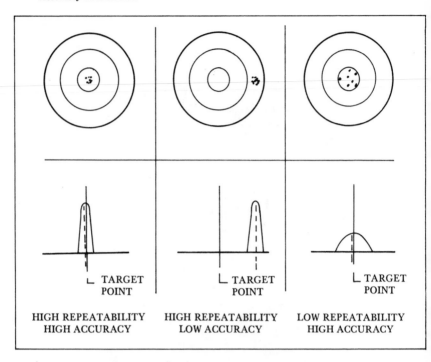

Figure 3.19 *The difference between accuracy and repeatability.*
If the target is the bullseye, then the top lefthand 'card' shows good
accuracy and repeatability since the shots are all clustered around the
target. With the top middle 'card' the clustered result shows good
repeatability, but the gross error (from the bullseye) shows poor accuracy.
The top righthand 'card' has the shots loosely clustered all around the
target, hence good accuracy but poor repeatability. The bottom three
illustrations indicate the distribution curves for the three 'cards'.
(from Verifying robot performance, *Robotics Today*, October 1983)

two points is further than a straight line joining them. Therefore the velocity stated in robot specifications should be used
as indicating that one robot is faster than another for performing a simple given action.

The selection of one robot over another for a given task is a
multifaceted procedure involving many interdependent factors. Ultimately, however, the selection is based upon the economically important parameter of cycle time.

CHAPTER 4
Calculation of cycle times

There are four methods of estimating the time to perform a task by a robotic system: simple, intermediate, most exact and parts tree. All are based on the fact that robot performance is predictable and repeatable, and that when the dynamics and control algorithms of a particular robot are known, it is a straightforward procedure to calculate assembly times.

Methods for calculating cycle time

SIMPLE METHOD
A given robot has an average time for performing average assembly tasks. If this time is t_c and the number of parts and tools used in the assembly is n, then the time required for assembly t_a is:

$$t_a = n * t_c$$

This assumes that the time taken to change a gripper or tool is the same as the average task time. If this is not so, then the equation can be written:

$$t_a = (n_p * t_p) + (n_g * t_g)$$

where n_p is the number of components, t_p is the average time to assemble a part, n_g is the number of grippers or tools used per cycle, and t_g is the average time to change a tool or gripper.
For a system with a variable number of grippers and robots, and assuming that any gripper is not changed more than once per cycle, the number of hand changes H can be calculated thus:

$$H = h - N$$

where h is the number of grippers and N is the number of robots. Therefore, the time taken for gripper changes t_h is:

$$t_h = (h - N)t_c$$

The approximate total time required for assembly, including gripper changes, will be the sum of t_a and t_h divided by the number of robots, assuming perfect efficiency in balancing the multirobot system. Experience shows that a 10% increase in overall time can be expected for each additional robot added. The total assembly time T can therefore be expressed as:

$$T = \frac{(t_a + t_h)}{N} + (N - 1)0.1 \frac{(t_a + t_h)}{N}$$

$$T = (0.9 + 0.1 N) * (n + h - N)t_c$$

As the value t_c is determined from averaging the time for many complete cycles, it is reasonable that inaccuracies will arise in the calculated value of T when the number of assembly actions per arm per cycle (n/N) is small. It is estimated that where the value of n/N is greater than five, then the calculated value of T, by this method, will be within 20% of the actual time.

The simple method assumes a task time of five seconds, irrespective of task, which for a single gripper/robot combination can be reduced to:

$$T = 1.0 \, n t_c$$

However, when the ratio of n/N is small, or the assembly sequence deviates significantly from a series of complete cycles, a more complex method of determining cycle time through a finer breakdown of the assembly cycle is used.

INTERMEDIATE METHOD
This more sophisticated method defines the basic task time as 8t, and can be extended or reduced according to the complexity of the approach and retreat motions required to complete the task. The pickup, discharge or use of any object requires an approach and a retreat of the robot's end effector from a point in the near vicinity of the object which is classified according to the following rules:

1. If the approach/retreat requires only one point-to-point move, no additional time will be added to the value of 8t.

2. For any action that requires a slow step, then an additional time 't' will be required.
3. The maximum additional time for any complete assembly cycle will be 4t.

The time required for the robot to move between one retreat and the next advance varies as a function of the distance travelled and the number of steps required. In addition, for a small complex action requiring less time than a longer simpler action, the task time 8t would be over generous for the smaller move. Therefore, for these more complicated motions, a time equal to 2t is deducted as follows:

1. For a single combined arm action, with the longest major motion being less than 0.3m.
2. Two combined arm actions using only minor axis motions that do not exceed half of the axis length, with no stop at any intermediate point.

For any complete assembly cycle, a maximum time deduction of 4t can be realized. Therefore, for a task time of 0.5 seconds, the total time for completed assembly can vary between the extremes shown below:

$$\text{Min} \qquad 8t - 2(2t) = 2s \qquad \text{to} \qquad 8t + 4(t) = 6s \qquad \text{Max}$$

There are assembly sequences that will not require a complete assembly cycle for each part. For example, one gripper could unload several moulded parts from a cavity mould before returning to the assembly fixture to deposit them. A half-task is therefore defined as moving from the completion of one retreat through the completion of the next, and is considered equal to 4t. This time can be modified using the same rules outlined above, with the maximum deviations allowed being limited to half of the previous values. Therefore, the time to complete a half-task will range between extremes:

$$\text{Min} \qquad 4t - 1(2t) = 1s \qquad \text{to} \qquad 4t + 2(2t) = 3s \qquad \text{Max}$$

Although this intermediate time system requires a very disciplined and logical methodology, it does result in a more accurate time estimate for each portion of the assembly task and permits balancing of a multirobot system.

An example of the intermediate time system is shown in Table 4.1, which was used to determine the time for assembling a transmission governor through the use of a double-armed

Unimate 6000 robot. The intermediate time system predicted an overall time of 30 seconds. The simple system predicted a time of 35.75 seconds and the actual time was 31 seconds. The value of 35.75 was obtained by using $n = 13$ (from the right-hand column of Table 4.1, which is the largest number of actions performed). Based on the equations given earlier $N = 2$, and $h = 2$.

MOST EXACT METHOD

Today's industrial robots often boast a life expectancy exceeding 40 000 hours, which is the equivalent of 20 shift years. Assuming that a robot does not become technically obsolete, this could mean that it will be performing a given assembly task for a number of years. To maximize efficiency and reduce costs, it may prove worthwhile to perform a detailed time analysis to predict cycle times accurately, especially when a balanced multirobot system is required.

An efficiency figure for a multirobot system consisting of N robots with individual assembly times of T_i seconds and resulting in an overall cycle time of T (including waiting time) can be expressed:

$$E = [\{\Sigma T_i\}/NT] 100$$

Using the values from Table 4.1:

$$E = \frac{(26.5 + 29)}{60} * 100 = 92.5\%$$

In developing multirobot systems, it is important to arrange the assembly sequence so as to minimize waiting time and thereby maximize the efficiency of the system.

This detailed time computation is an expanded version of the intermediate method, and requires a complete knowledge of the assembly station geometry including the relative positions of the robot(s), tooling, fixtures and parts, as well as a knowledge of robot dynamics and a detailed description of the robot control algorithms.

The arm trajectories for robots can be point-to-point and/ or continuous path. A generalized point-to-point step results in all articulations starting to move to their destinations, usually in unison. Each actuation goes through an acceleration, slew and deceleration period, with each motion often finishing quite independent from the others. Any number of these points can be linked together, with the condition that all articulations are in the true positional coincidence prior to the next step. Alternatively, points can be taught such that the memory will

left arm	time	units	right arm
SB off feeder and onto fixture $8t + t + t$	$10t$	$3t$	Insert two screws $4t - 2t + t$
		$6t$	Return screwdriver $4t + t + t$
LV off feeder $4t - 2t + t + t$	$3t$	$10t$	FA of fixture and onto exit chute $8t + t + t$
Align SB $4t - 2t + t + t$	$4t$		
Insert LV into SR $4t - 2t$	$2t$		
LV1 off feeder and insert into LV $8t - 2t - 2t + t + t + t$	$7t$	$4t$	LV3 off feeder $4t - 2t + t + t$
		$3t$	LV2 off feeder $4t - 2t + t$
SV off feeder and into SB $8t - 2t + t + t + t + t$	$10t$	$6t$	LB off feeder and onto fixture $8t - 2t$
		$4t$	Insert LV2 & LV3 in SB $4t - 2t + t + t$
SV1 off feeder and into SV $8t - 2t - 2t + t + t + t + t$	$8t$	$11t$	SV2 off feeder and into SB $8t + t + t + t$
Hold SV1 ($\equiv 3t$)		$6t$	Get screwdriver $4t + t + t$
SB off fixture and onto LB $8t - 2t + t + t + t$	$9t$		Wait ($\equiv 2t$)
Hold SB ($\equiv 4t$)		$5t$	Insert two screws $4t + t$

Table 4.1 *Assembly sequence with time estimates determined by intermediate time system*

update to the next step before position coincidence is met, thus creating a rounding effect which results in a significant reduction in activity time.

The time required to perform various trajectories can be calculated using information describing arm dynamics and control parameters. A continuous path trajectory can be programmed so that the end effector follows a predetermined path in space over a wide selection of velocities. The linear interpolation scheme used for continuous path steps allows a simple calculation of step velocity by dividing the number of interpolations per step i by the frequency of interpolations f.

The calculation of the point-to-point steps is more difficult and requires knowledge of the acceleration rates and slew speeds for each articulation. A typical velocity profile for a point-to-point step is shown in Figure 4.1, which portrays a linear acceleration a, followed by a slew velocity v (for steps long enough to saturate), finishing with a linear deceleration a. The calculation of the time required to move distance S is:

$$t_s = \frac{S}{v_s} + \frac{v_s}{a} \qquad S > \frac{(v_s)^2}{a}$$

The time required for a step of length S to be covered where velocity saturation is not reached is:

$$t_{ns} = 2\left[\frac{S}{a}\right]^{1/2} \qquad S < \frac{(v_s)^2}{a}$$

Because of the independence of completion times for each of the articulations in a point-to-point move, a generalized motion consisting of up to six degrees of freedom requires the calculation of the time needed for the slowest articulation. Knowing the coordinates $x, y, z, \theta, \beta, \delta$ of the points between which the motion will occur, the transformation matrix M of the robot enables the step length S_i of each articulation to be calculated. These values of S_i can be substituted for S in the previous two equations to determine t_s or t_{ns} respectively.

The time required for any generalized assembly task can be found by adding together all the times required for arm motions under normal memory control, associated time delays, waiting time and times required for all motions under sensory feedback.

The very nature of using sensory servoing (eg force and vision) to achieve part insertion will result in a variance of any predicted assembly time. Also programmed waiting time, which is defined as that time where any device (including

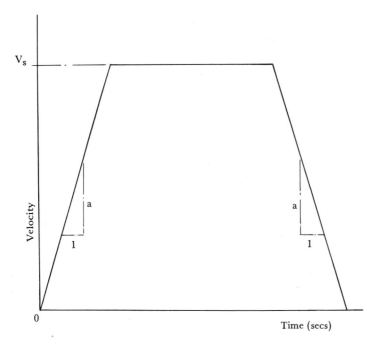

Figure 4.1 *The velocity profile of a robot.*
Often the slewing velocity V_S is not attained before the robot has to
begin decelerating to its terminal position.
(from *Flexible Assembly Systems*, Plenum Press, 1984)

another robot) causes the primary robot to be stationary, needs to be accounted for. This variable time can be attributed to external tooling (eg parts feeders, index tables, etc) or two-handed coordinated tasks where one robot may be used as a holding fixture. Programmed pauses do not present any problem since their duration is known and defined within the sequence.

The calculation of time required for a complete assembly system using this detailed method is tedious and needs a detailed layout of the assembly station, including the design of all the tooling and other ancillary equipment. Also a complete choreography of arm movements for the complete assembly cycle is needed. Using an assembly precedence diagram, a preferred assembly sequence can be developed for a system using a single robot, which can then be analysed for the overall cycle time.

Parts trees

Parts trees are an alternative to the mathematical derivation of a robot's cycle time, and are used to identify relationships and actions necessary to achieve assembly. Times for each action are usually allocated on an educated 'guesstimate' basis, and the total cycle time determined. The parts tree also contains information about any jigs and fixtures required, as well as data about each assembly task in terms of insertion depth and orientation.

Figure 4.2 shows the parts tree for the alternator shown in Figure 4.3. The tree is constructed by listing all the components and then drawing the 'branches' that relate each item to its subassembly. An orientation code is allocated to each item, which relates to the relationship between its 'presented' and installed position in the final assembly. For instance, a front bearing unit needs to be turned upside down before it can be inserted onto the rotor, hence code 1 is used for its components. On the other hand, code 0 is used for the fan, which has a unique assembly attitude (ie blades down) when presented in that attitude. Many components (eg washers or bearings) do not have any specific orientation, and are not coded.

The time axis of the parts tree runs from top to bottom, adding one component to another, or to a subassembly, which is indicated by an arrow. A node at the intersection of two lines indicates where assembly takes place. Lines connecting components already joined together have no arrow.

The orientation code for a particular component is placed close to these arrows, with a distance (in millimetres) representing the shortest path to the assembly. Values in brackets represent the distance a particular part can be dropped before coming to rest in the assembly.

For all such distances and activities, it is possible to estimate the time taken for each movement and thence for the whole assembly. The requirement of special fixtures is signified by F and the special operations (eg the aligning of parts, or the checking of functional assembly to ensure the rotor is turning freely) are specifically indicated.

Assessing workload

Before an assembly sequence can be arranged for a multirobot system, the workload for each robot must be carefully balanced to reduce costly waiting time. This requires the prediction of the time required by each robot to perform a complete

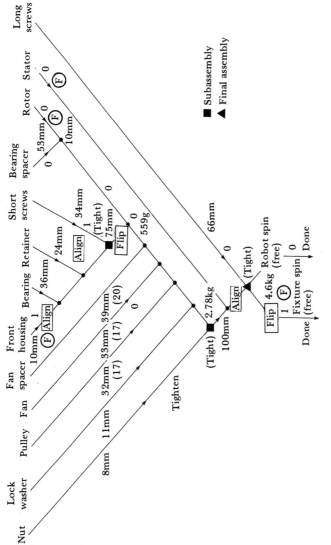

Figure 4.2 *A parts tree analysis of an alternator.*

(from *Design and Control of Adaptable-Programmable Assembly Systems*, by D. E. Whitney *et al.* NSF Grant No. DAR77-23712, 1980)

51

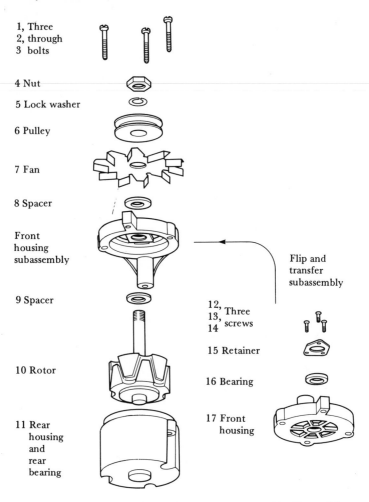

1, Three
2, through
3 bolts

4 Nut

5 Lock washer

6 Pulley

7 Fan

8 Spacer

Front
housing
subassembly

Flip and
transfer
subassembly

9 Spacer

12,
13, Three
14 screws

15 Retainer

10 Rotor

16 Bearing

11 Rear
housing
and
rear
bearing

17 Front
housing

Figure 4.3 *An exploded view of the alternator used in Figure 4.2.*
(from Figure 2-4 in *Design and Control of Adaptable-Programmable
Assembly Systems*, by D. E. Whitney *et al.*
NSF Grant No. DAR77-23712, 1980)

assembly task. It is also necessary to acknowledge that variations in cycle time can occur because of system down-time, due
to faulty components. These items, while incapable of being
processed, prevent system operation until they can be cleared.
An exception is the very sophisticated system, where each part
is checked before the next action is initiated. If a component is
found to be unacceptable, it is rejected and the action repeated
until a suitable part is presented, or the system closes down.

A robot can be used to manipulate items around the workstation and to monitor and control the assembly sequence, but it is the end effector that determines the efficiency with which these tasks are performed. Simple and intelligent end effectors are described in Chapter 5, along with devices that enable the robot to negate some of the vagaries of the workstation.

Grippers

In order that a robot can perform an assembly task, it must be fitted with a device often called the 'end effector'. The end effector can be either a tool (eg a sealant dispenser) or a grasping device commonly called a 'gripper'. In line with industrial practice, the term 'gripper' is used to mean a grasping device, tool or any other end effector.

Robots are usually specified without grippers, the reason being that grippers are peculiar to the task and the environment in which it performs. Consequently, grippers form part of the customizing package, alongside other auxiliary equipment, installation and commissioning. At present, there are two exceptions to this policy: the IBM 7565 and DEA's PRAGMA, both of which are supplied with grippers fitted with sophisticated sensing technology. Figure 5.1 shows the integral tactile and optical systems of the IBM 7565 gripper. The PRAGMA has a similar capability, which includes force sensing on its wrist.

Currently, there are no industrial standards for the fixing holes of end effectors, so every gripper must have a different set of mounting holes to suit each robot. Although a gripper can be configured to a specific task, the mechanisms used to operate them are similar. The grippers are then customized through 'add on' fingers (Figure 5.2).

It is usually assumed that a gripper has no independent degrees of freedom, as it is anticipated that all degrees of freedom are provided by other robot elements (eg the wrist and/or arm). It has been recognized, however, that certain devices (eg autoscrewdrivers) will, because of their inherent degrees of freedom, duplicate certain robotic capabilities.

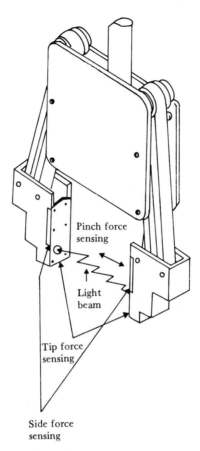

Pinch force
sensing

Light
beam

Tip force
sensing

Side force
sensing

Figure 5.1 *A robot gripper fitted with optical sensors and two sets of force sensors. Breaking the light beam indicates the presence of an object, whilst the force sensors can be used to detect presence, measure exerted pressure and for component detection.*
(from IBM 7565 Specification Booklet)

Sophisticated gripper versus simple gripper

The tasks performed during assembly often require precise movements of the objects being assembled. There are two alternative methods of achieving precise motions robotically:

1. To use a very sophisticated robot and a simple gripper, which has the advantage that the specialized equipment, namely the gripper, does not have to be technically complex, hence its inherent reliability will be high and its cost low. A disadvantage is that an

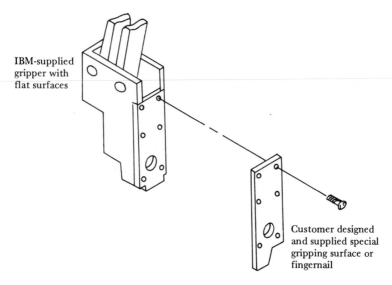

IBM-supplied
gripper with
flat surfaces

Customer designed
and supplied special
gripping surface or
fingernail

Figure 5.2 *A standard gripper mechanism can be customized through
the addition of fingers. Properly designed attachments still retain the
sensory capability of the basic gripper.*
(from IBM 7565 Specification Booklet)

expensive robot, which would include unnecessary
features, may have to be purchased. Additionally, there
may be problems in obtaining a robot with the
necessary intelligence and appropriate payload and
reach.

2. To use a simple robot and a sophisticated gripper. The
disadvantage here is that the gripper would have to be
designed specially for the task. Consequently, there
would be a long development period with a lot of
finance involved. Also, because it is specially designed,
its reliability would not have been tested through many
hours of field trials, resulting in a large risk element.
However, the advantages of this method are that a
lower priced robot may be used, and it is also probable
that the choice of units is enhanced since the needs are
structural rather than 'intellectual'. Simpler robots
have higher payloads. Therefore, the weight penalty of
the specially designed intelligent gripper will not
present a handling problem.

In addition to the technical merits of one robot-gripper
combination over another, the economic implications of the

decision must not be overlooked or ignored. (This aspect of robots for assembly is discussed in Chapter 15.)

Figure 5.3 shows an intelligent gripper (designed by Marconi Research Laboratories) for inserting integrated circuits with two to sixteen dual inline pins into printed circuit boards. The gripper can be reconfigured through computer control to match both the programmed input and to respond to any external feedback information on the component being handled.

Figure 5.3 *Intelligent gripper fitting a dual inline package to a printed circuit board. The finger blocks and thumb are clearly shown.*
© *1983 The General Electric Company plc.*
(courtesy of Marconi Research Centre)

Physically, the gripper is a multifingered device with eight pairs of 'fingers' and four 'thumbs'. The fingers are arranged in four blocks: one, two, four and one pair, respectively. Each block has one 'thumb' positioned normal to the fingers and parallel to the package's pins. The tips of each finger are electrically isolated from its support and wired so that electrical tests can be performed on the dual inline packages. The fingers are spaced 0.10 inches apart and are fitted with 0.10 inch diameter plastic optical fibres to position the package over the back-lit circuit board.

Three DC motors are used to configure the gripper, one activating the selected finger pairs, the second retracting the
remaining finger blocks outwards so they are clear of the work
zone, whilst the third controls the thumbs. Each finger block
has its own solenoid. Figure 5.4 shows how actuation of this
solenoid allows the fingers to be reconfigured. The reconfigur-
ation takes two to three seconds, but since this takes place
whilst the robot is moving to pick up the 'next' package, the
disruption to the assembly cycle is minimal.

Package insertion is performed solely by gripper motion,
with the robot (like that shown in Figure 1.1) acting as the
stable base.

The alignment of the package is achieved by an X, Y, Θ table
that supports the circuit board and moves in response to inputs
from the fibre optics fitted on the gripper. This maximizes the
light levels entering each fibre optic. The sensory system uses
4096 levels of light, which gives sufficient discrimination to
detect the centre of a hole. Insertion begins with a small extension of the thumbs (Figure 5.5), monitored to confirm that all
the pins are located in their holes without causing damage to
either package or circuit board.

Assuming this first stage has been achieved, the fingers are
opened and the thumbs extended to the desired insertion position. This action is checked by monitoring the current of the
drive motor against a preset limit known to be sufficient to
realize insertion. The final stage of the insertion procedure is to
check that a package has in fact been inserted, in case the
package had fallen from the gripper prior to the initial insertion. Thus, the thumb is extended at very low motor current
towards the surface of the circuit board; stalling of the motor
'confirms' that the package was present and has been correctly
inserted.

A gripper fitted to a simple robot must be able to compensate for the positional error between the desired target and the
range of repeatability of the robot, which in many cases could
be an order of magnitude. The gripper could include sensors to
identify the positions and orientations of fixed features in the
target object. These sensors could then be converted into error
signals and subsequently transformed into lateral and rotational motions of the gripper, which is in the horizontal plane
so the component is aligned with its mating part. The two items
are then joined by a vertical motion by either an integral part of

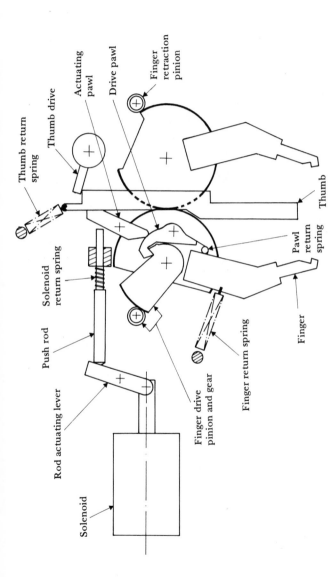

Figure 5.4 *Reconfigurable gripper mechanics. Mechanical construction of a one finger block. Operating the solenoid pushes the thumb across so that the thumb drive engages with the flat portion of its step. Simultaneously, the actuating pawl causes the engagement of the drive pawl and finger drive gear. The selected fingers are driven inwards until the finger retraction pinion is out of engagement with the finger cog. The unselected fingers are then driven outwards so as to clear the work zone.*
© *1983 The General Electric Company plc.*
(courtesy of Marconi Research Centre)

Figure 5.5 *Gripper showing the thumb pushing down on the top of the dual inline package.* © *The General Electric Company plc.*
(courtesy of Marconi Research Centre)

the gripper, or a gross robot motion, whichever is deemed appropriate.

Compliance
Irrespective of the intelligence of the gripper and/or robot, the assembly of two close fitting items is often fraught with problems. For instance:

1. Geometric error between the item being presented and the part into which it must fit often causes problems.
2. Errors occur through an inaccuracy in the robot control system.
3. Misalignment of the robot and gripper to the jigs and general workstation.
4. Manufacturing variations in the parts being assembled.
5. Differences in pickup positions.
6. Grasping attitudes of the component.

Bearing these problems in mind, an assembly task may have three outcomes: successful assembly, partial assembly ending with jammed parts, and collision of parts preventing assembly.

To maximize the number of successful outcomes, it is necessary that the robot-gripper compensates for all the known and anticipated errors of position. One option is the intelligent gripper (Figure 5.3), but this is an expensive solution to a complex problem. A simpler answer is to endow the gripper with some elasticity so that it can accommodate the errors and still achieve assembly. The solution is in two parts: (1) the design of the components and (2) the compliant device. The use of chamfers on both mating parts encourages the items to mate, since the target area is increased by a female chamfer and the initial 'clearance' increased through a male chamfer.

Over the past decade, the Massachusetts Institute of Technology has undertaken major research on the analysis and design of compliant mechanisms. There have been two approaches to the solution:

1. The use of passive devices that react to the forces and torques generated when a device is held on a flexible mount while attempting to engage another item. The reaction takes the form of compensating moments and lateral forces that try to zero those applied, and so achieve assembly.
2. Active compliance has been developed in Japan, which uses electromechanical devices to move the gripper so that the applied forces are zeroed out and assembly takes place. (In many ways, this is similar to the intelligent gripper shown in Figure 5.3.)

Figure 5.6 shows commercially available passive or remote centre compliance (RCC) devices. These use combinations of different elastomer pads, shims and geometry to meet the particular needs of the application. For very delicate applications, RCCs are available that use springs instead of elastomer pads.

Finally, due to their design and method of construction, many grippers have a 'built-in' compliance. However, this does not mean that compliance devices should be ignored, merely that the natural flexibility of the robot and gripper is sufficient for a number of assembly-related tasks.

Types of grippers
Although a gripper action is essentially one of opening or closing, both the design of action and the structure of the gripper

Figure 5.6 *Remote Centre Compliance (RCC) devices
from the Lord Corporation's Range.*
(courtesy of Metzeler & Lord Gimetall, Aldershot)

are very task specific, though not always product specific.
They have many actions:

1. Parallel motion.
2. Scissor motion.
3. One jaw may be fixed, the other moving.
4. They can be sprung open or closed.

Grippers can and do incorporate multiple mechanisms, but
broadly there are five classifications used:

1. Mechanical clamping.
2. Magnetic clamping.
3. Vacuum clamping.
4. Piercing grippers.
5. Adhesive grippers.

MECHANICAL CLAMPING
Mechanical clamping is the commonest mechanism, whereby
pneumatic or hydraulic operated devices apply a surface
pressure on a component. These grippers are available in four
different styles: parallel jaws, finger grippers, expansion/
contraction grippers, and anthropomorphic hands.

Parallel jaws that hold a component between flat and/or vee'd surfaces. These devices can have one or two moving jaws.

Finger grippers (parrot beak) either encase the component, or hold it at the very tip of the jaws.

Expansion/contraction grippers incorporate a flexible diaphragm, bladder or other device, which when activated expands or contracts so that a frictional force is applied to the component. This mechanism is of particular use when dealing with delicate components or those whose geometric shape precludes the use of the rigid clamping method.

Anthropomorphic hands (Figure 5.7) usually contain three fingers that pick up and manoeuvre items. Though mainly at the research phase, many examples have been shown to work.

MAGNETIC CLAMPING
Magnetic clamping uses electromagnetism to hold the component. Obviously, the system is only applicable to material that can be held by a magnetic force and to components and workstation environments that can withstand magnetic fields without incurring damage. One particular advantage is that, to a certain degree, it is component independent.

VACUUM CLAMPING
Vacuum clamping is the application of negative pressure to components so they 'adhere' to the gripper. The commonest style of vacuum gripper is the use of suction cups arranged in a pattern to suit the component, with the vacuum being generated by an ejector or vacuum pump. If the vacuum is continuously being drawn, the component to be moved will 'fly' up as the gripper approaches. This has two benefits:

1. The gripper does not make a 'hard' contact prior to lifting the component, so risk of marring is minimized.
2. Activity time is reduced.

PIERCING GRIPPERS
The piercing gripper punctures the component to lift it. The technique is only used where (slight) damage to the component is acceptable (ie clothing, etc).

Figure 5.7 *An example of an anthropomorphic hand designed for direct attachment to a robot. This particular configuration duplicates exactly the four fingers and thumb of the human hand.*
(courtesy of Cranfield Robotics and Automation Group (CRAG), Cranfield Institute of Technology, UK)

ADHESIVE GRIPPERS
Adhesive grippers are used for components that do not permit any of the above methods. The grippers make use of sticky tape to pick up and hold the component.

Gripper design

SIZE AND FUNCTION
The size and function of a gripper is directly related to dimensional size, material and weight of the components to be transported. The first parameter that defines the gripper design is static load, which defines the operational envelope as:

1. The component weight, in terms of the capacity of the gripper, determines the gripper's ability to accommodate the component. This limit is self-evident, and reflects the technical capabilities that have been designed into the robot and gripper.
2. The gripper's ability to restrain the component's weight by frictional and shear forces determines the weight that can be held without slippage under static conditions.
3. The attitude of the gripper jaws during a particular manoeuvre controls the weight that can be moved. If the jaws are always at right angles to the gravitational force, then the component's weight is withstood by frictional and/or shear forces; the frictional forces being those that act directly between the gripper jaws and the component surface. Shear forces come into effect when the component rests on top of the jaws so that the shear strength of both component and jaws resists the weight of the component. If the jaws are in any other attitude, then both the frictional and shear forces apply, but each to a lesser degree.

It should be remembered that the limiting force that can be applied to the component is one that will distort and/or cause damage. It is also necessary to ensure that the loading is uniform across the clamped surface, otherwise a local high spot could cause localized damage to the component.

MOVEMENT WITHIN A THREE DIMENSIONAL SPACE
This design parameter acknowledges that a gripper is used to move a component around in a three dimensional space. This movement generates dynamic forces on the component-gripper interface as the gripper accelerates and decelerates during its task performance. Two sources of inertial forces are generated: centrifugal and momentum. These inertial forces can lead to the bursting of materials, so the design of the component and gripper must be checked against this eventuality.

COMPONENT GEOMETRY
Geometry of the component is used to define the operational envelope. This refers to the maximum component dimensions that can be accommodated without affecting the desired per-

formance of the gripper. There are four conditions that cause instability of gripper performance due to component geometry:

1. When the component dimensions are such that the gripper cannot satisfactorily hold it.
2. Where one or more of the component's dimensions are such that the centre of gravity of the component causes a large tilting moment, either on the gripper or the components within it.
3. Where one or more of the component dimensions are such that the component cannot be manoeuvred without collision.
4. Where the component is so small that the gripper cannot function properly.

CONSTRUCTIONAL MATERIAL

A gripper is normally constructed of steel, which is a tough and easily obtainable material that is compatible with the majority of manufacturing processes. However, aluminium may be used when a nonmagnetic yet tough material is required. Like steel, aluminium is readily available and can be processed easily. Plastics are also used for constructing grippers to handle delicate components or when it is necessary to ensure electrical isolation. The use of soft materials (eg foam or rubber) is restricted to expansion/contraction type grippers, or as padding on the contact surfaces of grippers manufactured from more robust materials.

Ceramics may also be used to construct grippers used in hostile environments, such as those encountered in a forging operation.

Sensory control of grippers

Grippers can be simple grasping devices for moving items around a workplace, or intelligent units merely using a robot as a gross manipulator (eg the Marconi device shown in Figure 5.3). Between these two extremes, grippers can be fitted with sensors so that the robot is provided with tactical information prior to performing an action. These grippers are known as 'simple sensory grippers', which may be used to distinguish between components of different sizes before they are activated by the appropriate program to pick up a component. The sensors also enable a robot to function within a dynamic workstation by allowing it to check its sensors before every move to determine (1) what is happening, (2) if a component is available for pickup, (3) the identity of the component, and (4) if the other machines (if applicable) are ready for the process to continue.

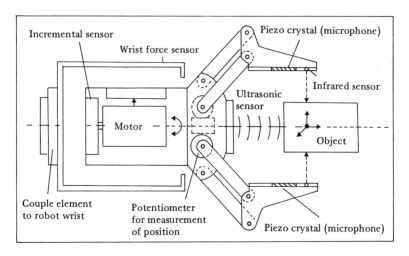

Figure 5.8 *A sensory controlled gripper.*
(from Institute for Information Technology at
Karlsruhe University, West Germany)

Figure 5.8 shows various types of sensors that *could* be fitted to a gripper. In reality, if the grippers are provided with sensors, they are limited to one or two types. For instance, grippers could be equipped with force sensors so the applied force is known and controlled or, alternatively, vision could be fitted within the gripper so that the robot can 'see' what is happening before it is activated (see Figure 5.9).

In essence the robot becomes an inspection system that measures, identifies and weighs the items prior to or during pickup. If the robot is servicing a number of different products, the gripper can be used to identify a particular product. If a product is identified as being correct the gripper switches the robot into the correct routine for that product, or if it is of poor quality or wrong size the product is rejected.

Gripper classifications
Grippers are either generally designed around a number of parameters (eg jaw opening and capacity) or specifically designed to handle one component and/or one task. The costs

Figure 5.9 *This assembly task involves fitting a 0.5mm diameter wire into a hole 0.7mm diameter. The position of the hole varies between one component and another. Therefore, before the wire is collected by the robot, a camera, fitted as an integral part of the gripper (see righthand corner), is used to search for the hole. The gripper rotates so the camera can scan the surface of the component until it finds the hole. The control system then determines the exact coordinates of that hole, taking into account the offset of the camera to the gripper faces. With the gripper in the insertion attitude, the wire is collected and placed into the hole.*
(courtesy of Dr John J. Hill, University of Hull, UK)

are inversely proportional to the amount of specialization, as is the reliability of the device. The time taken to perform a given 'complex' task would be less by a sophisticated specially designed gripper than a generalist gripper (see Figure 5.10).

A single gripper is used to perform a limited number of different tasks, since no universal grippers are available at present. If a wide number of tasks are to be robotized, then a number of 'gripper' options must be open to the user:

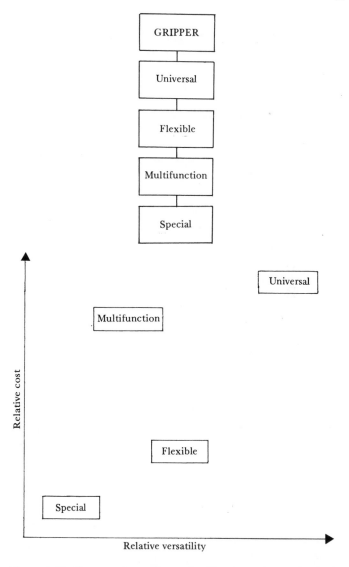

Figure 5.10 *There are four gripper classifications whose value and applicability are a function of their cost and versatility. The universal gripper has the highest versatility, but also has the highest cost since it would be difficult to manufacture a gripper to do 'everything'. The opposite extreme is the special gripper manufactured for a very specific task. Its relative cost is less than any of the others, but its versatility is virtually zero. All grippers can be fitted into a cost-versatility matrix, but the decision-maker must determine what is wanted and at what cost.*
(from *Flexible Assembly Systems*, Plenum Press, 1984)

1. Sufficient workstations must be made available to allow work to move to the 'next station' as the tasks for that station-gripper are completed.
2. Multiple robots can be used at one station, with each robot having its own gripper performing specific tasks.
3. Multiple arm robots.
4. One robot with multiple grippers.

Multiple robots

The PRAGMA assembly is a unique multiple robot in that up to four units are fitted onto a single rail which can then be operated by one controller to form one workstation. Fairey Automation (UK vendors of the PRAGMA assembly robot) have designed and installed a number of multiple robot systems. Figure 5.11 shows one such system, where the first robot (closest) places a plastic liner into a metal cup, and then places that cup onto a pallet that is transferred towards the second robot. Here, a tie rod is grasped by the robot and used to 'hook' off a ferrule from a vibratory feeder. The robot then uses its second gripper to pick up the cup processed by the previous robot, and transfers both 'assemblies' to a hydraulic press, where the tie rod and ferrule are forced into the cup. The third robot removes the assembly from the press to load into a testing machine.

Multiple arm robots

Figure 5.12 shows a double armed pneumatically operated manipulator, where the two arms are fixed by the user, through the use of mechanical clamps, to the desired angular and vertical relative displacement. The arm displacement is maintained as the manipulator rotates horizontally and lifts vertically. The only independent arm operations carried out are extension or retraction and gripper actuation.

The manipulator can be fitted with three arms and have three rotational stops. This way it is useful for transferring material through a multimachine process, where the machines and material feeder are arranged at a constant angular separation about the rotational centre of the manipulator.

Multiple grippers

Multiple grippers have two manifestations: multigrippers and discrete grippers.

Figure 5.11 *A three PRAGMA robot line assembling tie rods in South Wales. The robots are fitted to a single slideway that forms one edge of the workstation. Fairey Automation are one of a number of companies that market both robots and total turnkey robotic systems.*
(courtesy of Fairey Automation Ltd)

MULTIGRIPPERS

If the allowable time for a set of tasks is short, and changing grippers takes too long, a multigripper may be used (Figure 5.13). Individual grippers are fitted to a polygonal headed turret which is rotated so that the correct gripper is in position to perform its task. Obviously there are technical problems with the service interfaces, and ensuring that the grippers not being used do not collide with the product being processed. Tool/gripper size is a limiting factor because a wide range of sizes, or the use of 'very large' grippers/tools, means the robot has to withdraw a long way from the material being processed so the turret can be indexed.

71

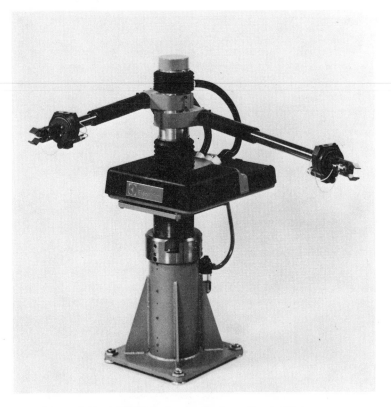

Figure 5.12 *A MHU Junior manipulator fitted with two arms.*
(courtesy of George Kuikka Ltd, Watford)

A good example of the use of multigrippers is well illustrated by Toshiba's robotized assembly of large electronic parts onto printed circuit boards at their plant in Fukaya, Japan.

Robotic assembly of printed circuit boards having small electronic components is not economically viable, since present day robots cannot compete with the existing well established technology of hard automation board populators. These are fed with ribbons of discrete components arranged in a sequence to match the assembly process.

It was therefore decided that a row of nine Tosman TSR-701H robots should be used to perform the assembly of the larger electronic components. The process rate is 1000 units per day and a cycle time of 20 seconds, which implies an operating efficiency of 75%. The design of the system (Figure 5.14) is in the form of three parallel lines, the first being a continuously moving conveyor that carries the printed circuit boards, the second a row of robots, behind which are the part feeders. The component feeding system has to cope with a large variation of different shapes and sizes (eg large condensers, transformers, terminal blocks, rheostats and integrated circuits in a standard multipin package, so components are presented to the robot's pickup position by pallet, gravity chute or vibratory bowl feeder). Each material feeding system is preloaded with approximately 1000 items, or enough to satisfy the demand for one unmanned shift.

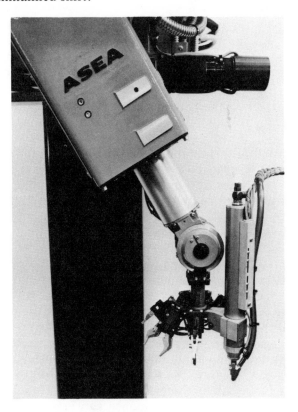

Figure 5.13 *This multigripper is specially developed for the ASEA 1000 robot, launched in October 1984. The gripper can be fitted with up to eight pairs of fingers, as well as the ASEA autoscrewdriver. The motions of the gripper are completely integrated and controlled by the robot's software and all programming is performed with the robot's programming unit.*
(courtesy of ASEA Robotics [Västerås])

The robot is provided with a multigripper, in the form of a drum (as illustrated in Figure 5.14), so as to be able to pick up a variety of components provided at its station. The drum's axis of rotation is parallel to the motion of the printed circuit boards, with the grippers mounted in a radial arrangement. When each gripper is needed, it is presented vertically downwards through the horizontal rotation of the drum.

The robot rotates on its base to the component presentation area behind, and uses one gripper to pick up a component from the 'first' feeder. It then indexes the gripper drum and moves to the second feeder, where the next component is picked up by its gripper. This priming of the gripper continues until all the grippers in that robot's drum are full. The robot then rotates back over the printed circuit board conveyor line and, using its gripper, inserts each component into the board in sequence.

Two elements that help the success of the system outlined above are the robot's positional repeatability and preciseness of location at the robot station of each printed circuit board. The robots have a repeatability of 0.05mm and the printed circuit boards are located with the same error. Component leads are within the same tolerance band of their true position and the holes in the printed circuit boards are within 0.10mm of their true positions. Together, these elements permit a successful insertion achievement of 99.9%. Control of the robots is through a master computer, which instructs each robot what program to use and when to start.

The assembly sequence begins with the insertion of three components, supplied from bowl feeders which lift the component to be processed to the precise height for the robot to grasp. At the second station, the robot inserts two more components, again collected from bowl feeders. The third station contains two gravity chutes and one bowl feeder, whilst the next uses bowl feeders. The fifth robot is supplied by gravity chutes and the sixth by three bowl feeders.

At the seventh robot, the components are presented on three pallets, each containing a number of parts for the robot to pick up in sequence. The eighth robot fits an angled plate supplied by the gravity feeder, whilst the ninth robot mounts the rheostat control. At the end of the line, the printed circuit boards are off-loaded and placed in groups of ten into racks.

These robots look as if they are of the arm and elbow configuration, but in fact they are closer to the SCARA. The inclined arm rotates through 300 arc degrees on the pillar to which it is mounted, while the elbow joint pivots on a vertical axis. The tool holder can also rotate and move vertically, just like the SCARA.

Figure 5.14 *Row of nine Tosman robots for the assembly line in Japan. The multigripper is seen quite clearly on the foremost robot. Between the arm and body of the same robot can be seen one of the many bowl feeders used to supply components to the robots.*
(courtesy of Evershed Robotics Ltd)

DISCRETE GRIPPERS

The cycle time implications of multigrippers caused by the need to withdraw, index and then reapproach the product, mean that discrete grippers often offer a more attractive option. In essence, the philosophy of these grippers is the same as tool changing in machining centres, in that the discrete grippers are stored on the robot workstation in a gripper/tool changing station (see Figure 5.15). However, time is needed for the robot to move to the change station, where it changes tools and then moves back to continue the task. With the discrete gripper, there are a number of technical problems that relate to the power systems being used by each tool/gripper. Because the devices are removed and refitted, the pneumatic/hydraulic/electric/electronic/optical/sensory circuits also have to be broken and rejoined. Although existing low technology de-

75

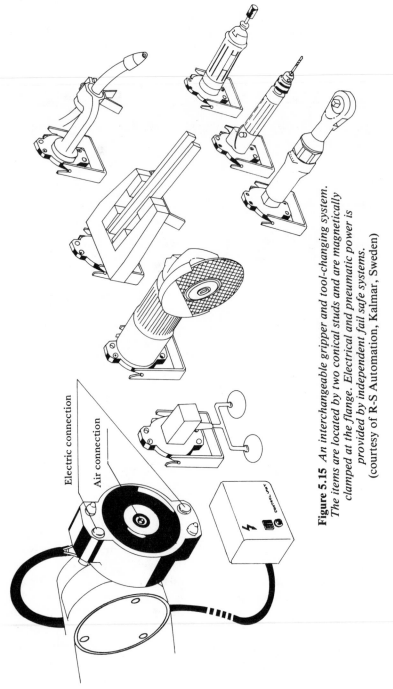

Figure 5.15 *An interchangeable gripper and tool-changing system. The items are located by two conical studs and are magnetically clamped at the flange. Electrical and pneumatic power is provided by independent fail safe systems.* (courtesy of R-S Automation, Kalmar, Sweden)

Electric connection

Air connection

vices allow interchange 'under pressure', the risk of an unsatis-factory reconnection is an additional, potential weak point in the system and a suitable length of time for the changeover procedure must be allowed.

CHAPTER 6
The assembly process

Fabrication and shaping of materials, followed by their joining and assembly into products, are the primary tasks within the manufacturing industry. Essentially, assembly is the fixing or forming of discrete items to and around each other. There can be very few sectors of the industry that do not make use of at least one joining process, though the processes used for this joining are very diverse. For example:

1. Welding used for joining large pressure vessels and the manufacture of miniature electronic components.
2. Mechanical fasteners used in the construction industry (bridges), as well as in precision instruments.
3. Adhesive bonding is at present used for a number of limited applications. However, with the increasing use of lightweight materials, honeycombs and plastics, structural adhesives will undoubtedly be used more in place of fasteners.

Kondoleon (1976) determined 12 tasks associated with the assembly process (see Table 6.1). These tasks include four (G, H, I and K) whose purpose is merely to assist an assembly process. The remainder are used in a variety of formats as assembly processes. It is indicative of the change in emphasis of assembly techniques since 1976, when Kondoleon did the analysis, that adhesives and integral plastic fastening were not included.

In addition to identifying the tasks used in assembly processes, Kondoleon also analysed their frequency of occurrence within a number of products and their components. From the analysis it is evident that simple 'peg in hole' and screwed assemblies are the most popular activities, with the vertically downwards attitude being the most common. Results of the analysis are shown in Figure 6.1.

Code	Description	Notes
A	Simple peg insertion	Location peg (slide/push fit)
B	Push and twist	Bayonet lock (quarter turn)
C	Multiple peg insertion	Electronic components
D	Insert peg and retainer	One component keyed by another (usually in a cross direction)
E	Screw	
F	Force fit	
G	Remove location peg	Reverse of A
H	Flip parts over	Between operation reorientation
I	Provide temporary support	
J	Crimp	
K	Remove temporary support	
L	Weld or solder	

Typical assembly related tasks were identified by taking apart and rebuilding a number of products and their components. Included in the analysis were a refrigerator compressor, an electric saw, an electric motor, a roaster-oven, a bicycle brake, and an automobile alternator. All the components and products could be assembled through various combinations of the above listed functions (after Kondoleon).

Table 6.1 *Categories of assembly tasks*

Assembly techniques

There are two distinct groups of assembly techniques:

1. Those that result in a permanent joint that requires the destruction of one of the elements to achieve separation.
2. Semi-permanent joints, whereby separation is achieved at will and without damage, assuming the correct tools are used.

As with everything, some assembly techniques could fit into both or neither groups. For instance, rivets can be drilled out, or have their heads sheared off; properly done, none of the elements held together by the rivets is damaged. Similarly, staples can be removed and the items separated without damage. However, the division used here is that if the joining medium (eg bolt, adhesive, pin, staple, etc) is reuseable after disassembly, then the assembly is said to be semi-permanent. Otherwise it is permanent (see Table 6.2).

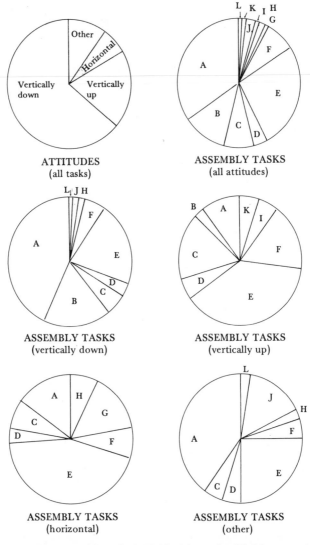

ATTITUDES (all tasks)

ASSEMBLY TASKS (all attitudes)

ASSEMBLY TASKS (vertically down)

ASSEMBLY TASKS (vertically up)

ASSEMBLY TASKS (horizontal)

ASSEMBLY TASKS (other)

Figure 6.1 *The assembly tasks in Table 6.1 are classified by two criteria: attitude of task performance and popularity of use. The top lefthand illustration shows the relationship between the four categories of attitude (ie vertically up, vertically down, horizontal [from any direction], and other). The top righthand pie diagram shows the popularity rating of the 12 assembly tasks for all attitudes. It can be seen that 'peg in hole' and 'screwing' outnumber all the assembly tasks by a very large factor. The relationship between a particular attitude and the assembly tasks is given in the lower four pie diagrams.*
(from Kondoleon 1976; source: *Flexible Assembly Systems*, Plenum Press, 1984)

Permanent	Semi-permanent
Adhesive/sealant	Bolting
Crimping	Clipping
Injected metal/plastic	Pinning
Soldering/brazing	Screwing
Staking	
Stapling	
Welding – Arc – Friction – Laser – Spot – Stud – Ultrasonic	

Table 6.2 *Examples of permanent and semi-permanent assembly*

Within these two assembly groups, two further divisions may be identified: (1) whether the robot is used *directly* to realize assembly, or (2) whether it is merely used as a device to load and unload the assembly machine. It must be acknowledged that membership of each group is made somewhat subjectively by the compiler and by the exact purpose of the assembly task.

The group using a robot as a handling, rather than an assembly, device is:

1. Bolting. This usually means 'screwed fasteners' larger than 6.00mm in diameter that often require nuts.
2. Riveting, with perhaps the exception of 'pop' rivets.
3. Soldering and brazing, though the application of the paste and components could be robotized.
4. Pinning, which is the placement of a dowel or sprung pin by an axial force.
5. Injected metal/plastic, in which discrete items are inserted into a moulding tool so that, at the end of the process, they form a single component. The injected material can be either plastic or zinc alloy.
6. Staking. The use of lanced tabs on a component which are inserts into the slot of another and then deformed by press operation.
7. Friction welding is achieved through relative rotation of two components about a common joint axis. Heat is

generated through the rotational friction and an axial force parallel to the joint face forges a solid state weld.

8. Ultrasonic welding is the joining together of items through vibrations that cause material flow.

The group using a robot to perform the assembly task directly is:

1. Adhesives and sealants, dispensed as blobs or beads and used to join an ever-increasing range of materials. The difference between them is nebulous, since they both stick and seal.

2. Screws are usually threaded devices less than 6.00mm in diameter and are available as either self tapping, in which the screw cuts its own thread into a drilled or punched hole, or standard threaded which require a tapped hole into which the screw is threaded. Both types of screws can be robotically inserted with auto-screwdrivers.

3. Crimping is the deformation of one component into a hole or groove of another. It requires a large force to do this, so a special gripper/tool is normally fitted as the end effector of the robot.

4. Clipping is the 'snapping' together of two components, after elastic deformation of one or both units. All that is required to achieve assembly is a light axial force and precise location of the two items being assembled.

5. Stapling and its close associate stitching are used for highspeed fastening, or where predrilled or prepunched holes are not desired. Of the two techniques, stapling is the easier to perform robotically, by fitment of a stapling gun at the robot's end effector. Stitching is more likely to be performed by a machine through which the material is fed. However, as in many such cases, the division between the two techniques is not easily defined since close stapling could be called stitching, and *vice versa*.

6. Welding (arc, laser, spot and stud) are all compatible with robots through the fitment of the equipment as the end effector of the robot.

Tasks can be graded into degrees of difficulty, based upon the sophistication of the robot control system needed to achieve assembly. For instance, the least difficult would only need a 'coarse' pick and place device, whereas arc welding requires the most sophisticated use of sensor technology in order to realize the desired quality of assembly.

The value and use of assembly techniques within a robotized assembly are unfortunately very subjective, mainly because designers and industries are often reluctant to leave a technique that has proved successful 'for generations'. However, the following assembly techniques are known to be satisfactory when linked with robots:

1. Dispensing adhesives.
2. Combination of welding gun and robot.
3. Ultrasonic welding.
4. Stud welding.
5. Autoscrewdrivers.

It should be recognized that assembly is much more than the techniques themselves. It includes the movement and correct positioning of items around a workplace, together with the design of the product and the process that allows the assembly to take place with the minimum of problems and at maximum productivity.

DISPENSING ADHESIVES

Robots can obviously be used for dispensing adhesives as they enable the correct amount of material to be put in the right place, instead of randomly at approximate positions, as is the case with manual application, when often too much adhesive is used.

Figure 6.2 shows a sealant dispensing 'robot', which does not look like any of the configurations outlined in Chapter 3. However, it is a robot on the basis that it is both computer controlled and easily reprogrammable by computer techniques. The pattern is entered by offline programming onto an EPROM chip, which is then used with component recognition sensors to control the dispensing of sealant.

COMBINATION OF WELDING GUN AND ROBOT

The combination of the welding gun and robot is well proven, with many hundreds of applications in both spot and arc welding. Laser welding used in conjunction with robots is becoming more common, especially with the newer and more compact higher energy lasers.

Figure 6.2 *This sealant dispensing robot may not be everyone's idea of a robot. However, it is an easily programmable and computer controlled device for manipulating tools through variable programmed motions.*
(courtesy of VS Remek Ltd)

ULTRASONIC WELDING

Ultrasonic welding has long been associated with robots, though the process often uses the robot to load the items into the welder rather than fitting one portion of the welder as the end effector of the robot.

In 1983, a United States Robots' Maker 100 robot (Figure 6.3) was installed at Amprobe Instruments Division of Core Industries Inc (Kehoe 1984), to ultrasonic weld two brass inserts into a volt-ohm-amp meter case. The cases are gravity fed to a pickup station from where they are individually transferred to the ultrasonic welder. The robot then collects the two inserts from a vibratory bowl feeder and places them into their respective case cavities. An output signal from the robot starts the welding cycle that melts the inserts into the case.

STUD WELDING

Stud welding involves the placement of a stud and the induction of the welding action through pressure and electrical power. Autofeeding of the studs to the welding head is proven technology and, since the arrangement of the studs is on a horizontal matrix, the use of a robot to move the welding head into

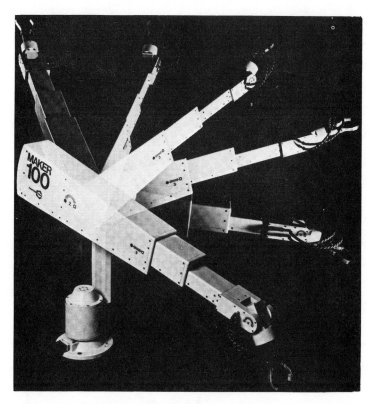

Figure 6.3 *The Maker robot, marketed in the States, has a virtually spherical work envelope of 1100mm radius, and a payload of 2.3kg.* (courtesy of United States Robots)

the correct position is very simple. The only robotic criterion is that the robot be strong enough to carry the welder and withstand the reaction force of the welding.

AUTOSCREWDRIVERS

Autoscrewdrivers are used in many robotic assembly applications. An example, shown in Figure 6.4, uses a robot fitted with an autoscrewdriver robot to assemble sheetmetal heat exchangers.

Figure 6.4 *An autoscrewdriver, fitted as the end effector of a robot, is used to assemble heat exchangers. The heat exchanger consists of two outer panels and two inner brackets. The robot screws one panel to the two brackets, after which the subassembly is manually turned over. The product is finished by fitting the second panel to the subassembly. In all, 32 self-tapping screws are used.*
(courtesy of VS Remek Ltd)

CHAPTER 7

Product and process design for assembly

The efficiency with which a robot will perform assembly tasks is directly related to the design of the product being assembled and the manner in which the processes are applied. Products are designed according to five criteria:

1. The individual components can be manufactured on appropriate machines at the lowest cost. For instance, designing for the best process related position for the parting line on moulded items, obtaining least waste on parts blanked from sheet material, and ensuring that other parts are designed for easy manufacture from bar or rod.

2. The cosmetic appearance of the product. This usually only pertains to those items that form the exterior surface or fittings of the product. The need here is for smooth, undamaged and unmarred surfaces that could be in conflict with certain manufacturing criteria (eg the placement of a moulding's parting line).

3. Safety, which has to be considered in the design of many products (see Chapter 12). A design is controlled by various codes and regulations that define closeness of items, maximum size of apertures, routeing of cables, and the need for internal partitions.

4. 'Functionability', in that the product must be structurally strong enough for its designated use and contain elements that have been chosen for their technical compatibility. In reality this means that the product design has a number of 'built in' constraints that take precedence over the above mentioned criteria.

5. Method of assembly. Apart from products designed specifically for hard automation processing, the method of assembly seems to be approached from the

basis of how a person would perform the task and with what dexterity and flexibility. Traditionally, assembly processes have been valued least and considered last, mainly because labour was cheap and plentiful. Nowadays, the need for products to be designed for automated assembly is essential for a company's survival. The reason for this is because a product's saleability is based upon its quality and price, both of which can be controlled more effectively by robots.

Product design is not enough to ensure that a robotic assembly system reaches its targeted productivity. Chapter 8 discusses the design of the workstation and its influence upon the assembly system as a whole.

Product compatibility

A product designed for manual assembly cannot be assembled robotically without redesign. If a company installs a robot to assemble an existing (manually processed) product without investigating the technical implications, and without implementing some, if not all, of the desirable modifications to product and/or tooling, the result will, at the very least, be an expensive and inefficient process. At worst, it could be a total disaster, with the company losing the market for that product and souring them from ever using robots again.

There are two sources of products for assembly by robot:

1. New products that exist as concepts, or redesigned versions of existing though outdated products.
2. Existing products that are deemed to have an expanding market and output and where the manual process cannot match the increased demand.

Another reason for the robotized assembly of an existing product is when management has proved that the manual performance of a process is directly responsible for high scrap rates, or poor and inconsistent quality. Either way, the method of evaluating a design for compatibility with robots is the same. The only additional considerations are incurred costs due to any changes to the design of components or processes (eg stocks of existing components being scrapped), and/or the need for compatibility with products that have or are being assembled manually because of servicing, stocking or approval reasons.

Method of construction

Each product and subassembly can be described by one of four classifications:

1. Frame construction (Figure 7.1). Items such as television sets or computers are frame constructed as they need a frame onto which other items can be mounted.
2. Stacked construction (Figure 7.2). Designs that require the components to be assembled one on top of another (eg armatures).
3. Base component construction (Figure 7.3). Items that incorporate a base onto which all components are fitted and transported through the assembly process (eg printed circuit boards).
4. Modular (Figure 7.4), in which individual sub-assemblies are combined to form different products (eg manufacture of automobiles in which different combinations of similar 'options' are used to produce a wide range of models).

Products may be grouped according to these classifications so as to determine their suitability for any given robotic work-station. Alternatively, if it will aid automated assembly, they can be redesigned (where appropriate) into another classification.

Fifteen design rules

There are 15 key principles (see below) to observe when analysing a product for assembly by robot. Many of these design principles, whilst being correct and valid, do themselves form composite base rules that should be acknowledged by product designers. An important rule is to acknowledge that a component's shape has some technical implication upon the assembly process. When assembled into a product, the surfaces of each component are usually categorized as either functional (ie if they are used for some purpose) or nonfunctional (ie if they do not, for instance, satisfy some form of datum or clamping need).

As items are being processed, their shape may change from a functional to a nonfunctional surface, or *vice versa*. It is therefore necessary for the product designer to understand, in depth, the manufacturing and assembly process when the components are being dimensioned:

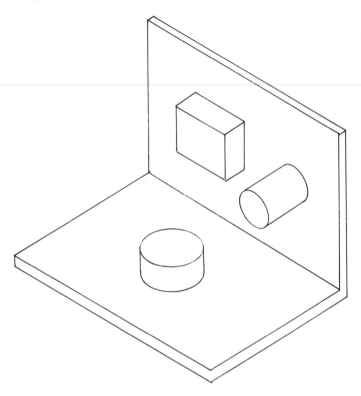

Figure 7.1 *Frame construction is identified when discrete items are fitted to a three-dimensional larger object.*

1. The robotic implications are that surfaces can be identified (at the design stage) as 'gripping surfaces', which minimize the risk of damage by the gripper to a (product) functional surface.
2. If the gripping surface is properly toleranced, then the risk of placement error, inherent with objects specified with wide manufacturing tolerances, is minimized.

Therefore, proper consideration of the function of each component in its progress towards total assembly will ensure that in-progress damage is prevented, and repetitive action is without risk of placement error and/or collision.

FIRST DESIGN PRINCIPLE
It is essential to check that every item within an assembly is necessary, since often components are included because of invalid or historical reasons. An objective view is often able to identify

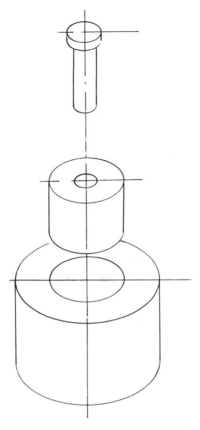

Figure 7.2 *Stacked construction refers to a group of items
that are assembled along one axis. Often, but not exclusively,
the components are symmetrical.*

a number of redundant components and/or suggestions where several individual items may be replaced by one composite component.

An important example in the use of this principle is the robotization of a group of products. Often, a company manufactures ranges of products whose basic construction is identical. For example, electric motors are all stacked construction, include similar, if not identical, components and are all assembled in the same sequence. If such a range were to be assembled by robots, then it would make both economic and technical sense to evaluate all the components and processes used, and attempt to minimize the numbers of discrete components needed to be handled. Obviously, certain components would

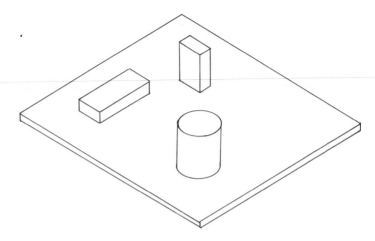

Figure 7.3 *Base construction is identified when items are fitted to an essentially two-dimensional larger object.*

be peculiar to a particular member of the range, where others could be reconfigured to form a common item across the range.

Redford and Swift (1980) analysed items contained in an immersion heater control unit. Figure 7.5 shows the original design and Figure 7.6 the same unit after analysis. The part count was reduced from 23 to 16 and 61% of the components in the new design were suitable for robotic assembly, compared to only 33% in the former design. As designed, the unit contained two terminal strips (blades), one of which is a carry over from a previous two pole model which served no purpose. Figures 7.7 and 7.8 show the control unit before and after analysis.

SECOND DESIGN PRINCIPLE

A precedence diagram should be developed and used to check the sequence of assembly. This often highlights any redundant assembly movements (eg going twice to the same bowl feeder in a single cycle). The amalgamation of components can be indicated by a 'node' that has a number of separate items fitting together. Balancing of multistation assembly stations may be determined from the diagram by comparing the cycle times of individual nodes. To minimize variation between nodes, tasks may be modified at each node.

A precedence diagram will also indicate whether or not it is possible to assemble parts incorrectly (in attitude or sequence)

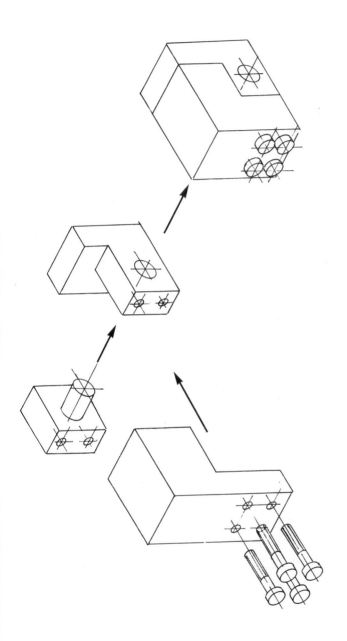

Figure 7.4 *Modular construction where a number of discrete subassemblies are assembled into a single product or major assembly.*

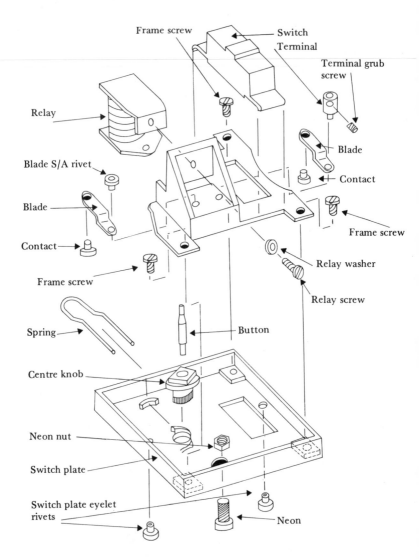

Figure 7.5 *Exploded view of immersion heater controller before a rationalization study into its functional and design requirement.*
(courtesy of Dr Alan Redford, University of Salford, UK
and Dr Ken Swift, University of Hull, UK)

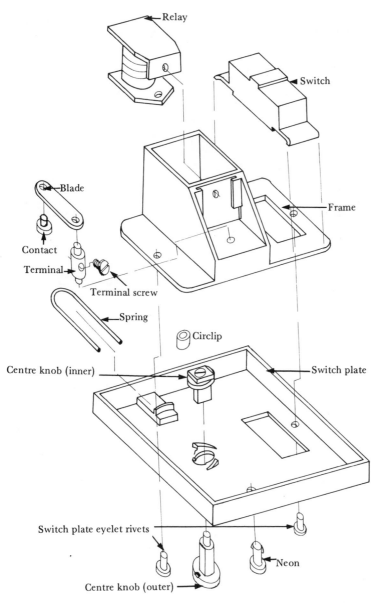

Figure 7.6 *Exploded view of the immersion heater controller after a study. Even a casual comparison with Figure 7.5 shows how much easier the 'new' model is to assemble automatically.* (courtesy of Dr Alan Redford, University of Salford, UK and Dr Ken Swift, University of Hull, UK)

Figure 7.7 *Immersion heater controller before redesign.*
(courtesy of Dr Alan Redford, University of Salford, UK
and Dr Ken Swift, University of Hull, UK)

by indicating whether the item is symmetrical or asymmetrical. The possibility of achieving a 'finished' product without using all the parts is also indicated by analysis of the precedence diagram, where it may be seen that the omission of one screw from a set of three because the autoscrewdriver does not feed in time or has run out of screws; or the omission of a bearing seal because a bowl feeder jams or a gravity magazine is empty, will neither drastically affect the assembly process or be obvious; the parts retained by the screws will still be held and the shaft fitted to the bearing will still rotate. The precedence diagram indicates those items and tasks where problems *could* occur so, in turn, system designers can then include inspection routines after each assembly activity that will check the assembly was performed correctly.

THIRD DESIGN PRINCIPLE
It is important to ensure that each component is fully and correctly specified. Robots accept all presented items in 'good faith', because they do not (as yet) have the capability to distinguish between good and bad parts. Of course, this sweeping

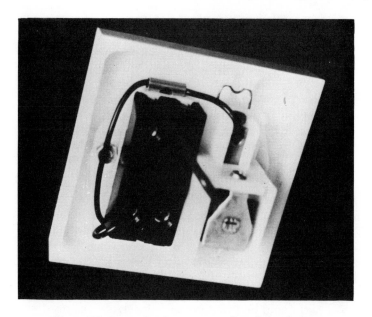

Figure 7.8 *Immersion heater controller after redesign.*
(courtesy of Dr Alan Redford, University of Salford, UK
and Dr Ken Swift, University of Hull, UK)

statement ignores gross discrepancies that can be detected by
bowl feeders or vision systems. The variation of components
from their specifications occurs because assembly systems are
built on the understanding that the components and products
being processed are in accordance with their manufacturing
drawings. In reality, the manufacturing department often
relax the tolerance band for acceptable components because
they know that the particular dimension has no functional
value in the operation of the product. Alternatively, a batch of
parts may have a deviation that is accepted by the management
so as to avoid the cost of rework or scrapping the entire batch.
Unfortunately, in both cases, this decision is usually made
without any consideration of the implications on the assembly
process. To overcome these difficulties, there are a number of
options:

1. Find out the specification to which the parts are
   presently being made, since the drawings used for the
   system design may be out of date. Agreement should be
   reached on the description or revision level that
   correctly identifies the manufacturing quality of those

items. If the specifications are known and agreed upon
many problems will never occur.

2. Ensure that the assembly process makes use of only the
functional dimensions of the components which are less
likely to liberal interpretation within the manufacturing
sector of the company.

3. Make the robot assembly tasks as 'fluid' as possible.
This is achieved by using self-centring grippers,
kinematic jigs and a lot of compliance in the robot-
gripper interface.

It is acknowledged that robots are used to perform selective
assembly routines. For instance, if a part does not fit, it is
rejected and another selected until the assembly task is realized. However, even these systems require components to be
within acceptable and known tolerances, since it is the variation between tolerance bands that identifies the component
being 'inserted'. If, for example, two components had similar
tolerances, but their deviation was gross, it is possible that the
wrong part could be fitted because the errors on specification
allowed it to happen.

For each item being assembled, there are five specification
groups:

1. Dimension, which ensures that components are within
dimensional tolerances specified by the drawing,
catalogue or data sheet.

2. Function, where a check is made that components
operate (ie move, switch, open, transmit or receive) in
accordance with the specifications.

3. Quality relates to the 'grade' (ie military or
commercial) and condition of the components (ie the
number of statistical quality checks and/or the
'appearance' of the part).

4. Material specification should be specific, not general
(ie it should give the national standard number, alone
or as a minimum, rather than just the group name of
'steel', 'aluminium', 'wood', 'plastic', etc).

5. Shape can be very important on the interfaces of
mating parts. It is controlled by geometric tolerancing.
For instance, it can be of particular importance when
using automatic printed board populators in which the
component leads need to be within a tight tolerance
band of straightness and/or pitch. Leads that are bent

and/or twisted will not enter the holes on the printed circuit board so the board would have to be retrofitted before or after automatic soldering.

FOURTH DESIGN PRINCIPLE

Any component variation must be minimized. For instance, within any assembly there are often a number of similar items (eg screws) and if these were all made the same, a considerable saving would be made in storage space and costs, in economies of scale, in additional equipment (eg bowl feeders and other assembly or handling equipment), as well as resulting in a more efficient and quicker assembly cycle, as the example below shows.

In Japan, Nippondenso (Naruki 1981) manufacture a wide range of gauges for the automobile industry. Historically, they had been made to each customer's specification, though analysis showed that functionally many of the items stocked for manufacture were similar and that rationalization of the components used could yield savings to both the manufacturer and the user. Figure 7.9 shows the construction of the gauge; the 'visible portion' remains customer specific because this is what the driver sees, whilst the 'hidden portion' was deemed suitable for rationalization. Figure 7.10 shows the part count before and after this analysis, which yielded a reduction of the types of components needed from 48 to 17. Mathematically, from the remaining components, it is possible to assemble 288 different products.

FIFTH DESIGN PRINCIPLE

Symmetrical components must be used wherever possible. This will reduce handling, orientation, identification and presentation problems. If orientation of a component can be correctly obtained by direct transfer from any one of its natural resting positions, it is obviously less expensive in time and equipment than having to orientate and arrange the component before the assembly task is accomplished. Many asymmetric components can be made symmetrical by adding nonfunctional features which add little or nothing to the cost of manufacture. Figure 7.11 shows a simple and common symmetry problem, ie pegs with different ends. Only one end functionally needs to be spherical, yet technically there is no problem making both ends the same and avoiding any need for orientation.

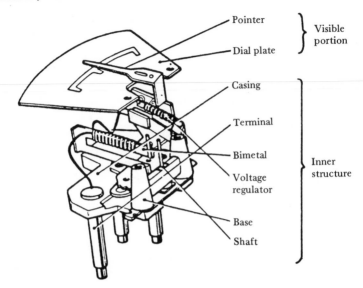

Figure 7.9 *Construction of a typical gauge, showing the individual parts as well as the dividing line between those that can be standardized (inner structure) and those that are customized (visible portion).*
(from Nakuki 1981)

SIXTH DESIGN PRINCIPLE
This design principle is the opposite of the fifth, namely if a part has to be asymmetric, this asymmetry should be exaggerated so that it is obvious to the feeding and handling equipment. In this way, before assembly, the presentation, attitude and orientation of the part can be detected and checked by crude but effective devices. Figure 7.12 shows a component made asymmetric by two diametrical holes. To avoid the use of complex detection systems, a non (product) functional slot is added. A simple guide rail will then automatically orientate the component.

SEVENTH DESIGN PRINCIPLE
The number of separate parts used in a product must be minimized. In many cases, screws and washers are supplied separately (for manual assembly). This means that each part has to have its own feeding and handling equipment, and the number of assembly tasks is increased. An easy answer to the problem is to combine some of the items by, for example, using captive washers on the screws. Therefore the number of pieces of feeding equipment (for those items) would be halved, though it

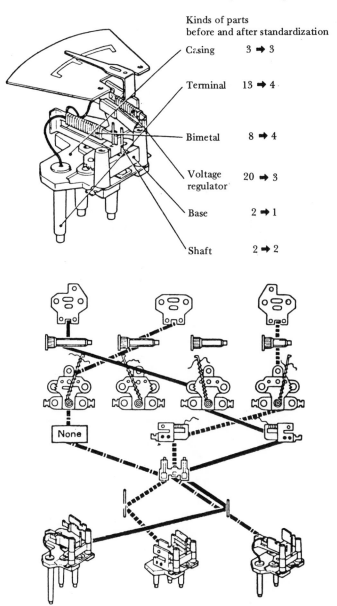

Kinds of parts
before and after standardization

Casing   3 ➡ 3

Terminal   13 ➡ 4

Bimetal   8 ➡ 4

Voltage   20 ➡ 3
regulator

Base   2 ➡ 1

Shaft   2 ➡ 2

Figure 7.10 *After standardization and rationalization of the inner structure of the gauge, the number of parts used for a range of gauges is reduced from 48 to 17 components.*
(from Nakuki 1981)

101

Figure 7.11 *(a) Asymmetric item that is difficult to orientate.*
(b) Symmetrical item manufactured through the addition of a (product)
redundant spherical end. The net result is that automatic feeding
of this item has no technical problems.
(from Designing for automated assembly, *Engineering*, June 1983)

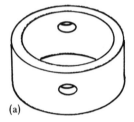 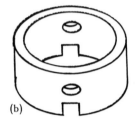

Figure 7.12 *(a) Component is asymmetrical, and presents orientation*
problems that cannot be resolved simply. (b) If a (product) nonfunctional
slot is added, the component can be simply and easily orientated
through the use of a slide rail on the feeder.
(from Designing for automated assembly, *Engineering*, June 1983)

might be more expensive per piece of equipment. There is also a corresponding reduction in the number of assembly tasks.

Another variation of this principle is to use adhesives or 'snap in' plastic legs to achieve assembly. Both methods are used in place of traditional fasteners (eg bolts, screws, dowels or rivets) which require the costly manufacture of a hole, which is then filled with an expensive item.

EIGHTH DESIGN PRINCIPLE

It is necessary to ensure that the component is presented at a known and consistent rate to the robot pickup station by the feeder device. Unless items are specifically designed for automatic processing, problems of tangling, nesting, telescoping and shingling are likely to occur:

1. Figure 7.13 shows three types of components that can be functionally 'reduced' to hooks and eyes to engage into each other. Tangling is avoided if the design concepts shown in Figure 7.13 are followed.
2. Figure 7.14 shows a plain conical item. Adjacent items to this design will nest and lock together. To avoid this a stepped inner shoulder is used to prevent locking of

WILL TANGLE   WILL NOT TANGLE

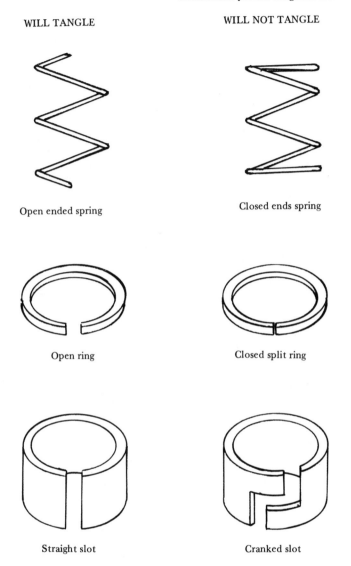

Open ended spring   Closed ends spring

Open ring   Closed split ring

Straight slot   Cranked slot

Figure 7.13 *In the lefthand column, these simple components will tangle since they each have hooks that will engage in adjacent components. The simple solutions are shown in the righthand column. The closed spring is a common occurrence and the closed split ring will still permit assembly to another component through radial pressure at the joint line. The third example is perhaps more radical, but under certain conditions could be cost effective.*
(from Designing for automated assembly, *Engineering*, June 1983)

the cones and limit nesting to a known and fixed distance, whereby items can be mechanically separated.

3. When items telescope, their degrees of freedom are reduced to that of forward along the axis (out of engagement) and rotation about that axis. This means that the engaged components cannot be manipulated into any other attitude or orientation and a log jam occurs. Figure 7.15 shows an example of a component containing both an axial hole and a stud. This hole is the same diameter, or bigger than the stud so that the stud can easily enter the hole, thus telescoping the individual items. To avoid telescoping, the hole should be smaller than the stud.

4. The vibratory motion of a bowl feeder 'encourages' components to ride up onto each other as they are moved along the track. This shingling (Figure 7.16) is prevented by thickening the 'contact walls' of the components and/or fitting a roof over the feed track so that they cannot mount adjacent items.

Will nest   Will not nest

Figure 7.14 *The lefthand component will nest, but the solution is simple – redesign with an internal step to prevent the taper from engaging and locking.*
(from Designing for automated assembly, *Engineering*, June 1983)

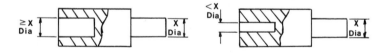

Will telescope   Will not telescope

Figure 7.15 *The lefthand component will allow telescoping because its bore is the same size or larger than the stud end. The remedy is a redimensioning of the bore to make it smaller than the stud.*
(from Designing for automated assembly, *Engineering*, June 1983)

NINTH DESIGN PRINCIPLE

Ideally, a product and its components should be designed for unidirectional assembly so that the components can be simply placed one on top of each other. This makes technical and economic sense when compared with products that require contorted gymnastics before completion can be achieved. A new development that epitomizes this principle is the use of surface mounted integrated circuits (ICs), which are laid onto and not through terminal pads. This overcomes any problems of twisted or bent IC leads, and allows a faster populating rate.

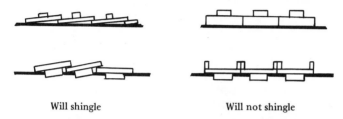

Will shingle   Will not shingle

Figure 7.16 *The two lefthand illustrations will permit shingling. One solution (shown on the right) is to thicken up the contact face.*
(from Designing for automated assembly, *Engineering*, June 1983)

Another consideration is shown in Figure 7.17, where an item has its surface features duplicated so that the feeding device does not have to identify the 'correct side' and, when necessary, turn a component over.

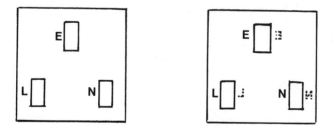

Figure 7.17 *The lefthand item has to be orientated the right way up, whereas the righthand component can be presented either way up, because it has the information on both sides.*
(from Designing for automated assembly, *Engineering*, June 1983)

TENTH DESIGN PRINCIPLE

The number of assembly functions performed by each component should be maximized so that the number of individual parts within the product is minimized. For example, electrical wires could be clamped in place between the two halves of the plug, instead of being screwed or soldered to pins which are themselves then clamped in position. Alternatively, wires could be 'moulded' into position with injected plastic. In both cases, elimination of the screwing/soldering operation reduces assembly and component costs, as well as increasing production rate.

ELEVENTH DESIGN PRINCIPLE

The product must be designed to suit machine work ethics and principles, rather than be designed to suit human methodologies. It should be determined how best a robot could handle the parts, and what assembly sequence would be preferable. Also, it is important to ensure that all the parts can be automatically presented in the correct attitude and orientation, at a rate acceptable to the system. In essence, this means ensuring that assembly is achieved through simple pick and place motions, and that the product and components are designed so that simple jigging is used to position the parts accurately.

TWELFTH DESIGN PRINCIPLE

The orientation of a subassembly must remain known and preferably constant throughout the assembly sequence. In other words, many components are supplied randomly, in bulk, to the workstations where they are ordered through a number of devices prior to processing. Naturally it is sensible, logical and economical to ensure that, once control has been gained over a chaotic situation, it is not then lost through random 'dumping' of processed subassemblies and products into containers.

This design principle is peculiar to multistation assembly processes in which subassemblies are passed directly through a number of assembly stations, or where a subassembly is built up to a particular level in one work area and then sent to a warehouse until required by another area for further work or completion into the final product.

THIRTEENTH DESIGN PRINCIPLE

This principle is associated with the twelfth in that, when a sub-assembly is moved, it should be structurally sound, or protected by its jig, so that it can be handled and moved without risk of damage. There is little point in performing an assembly task if items fall apart during transit between workstations. This assumes that both the transit medium and the build level of the product are compatible, and that deliberate upsetting forces are not imposed.

FOURTEENTH DESIGN PRINCIPLE

A subassembly should not be committed to a particular product until it is as far up the assembly chain as possible. Following the fourth and seventh principles of minimization of variability and number of parts, a given stage in the assembly sequence of a subassembly could be used for any one of a number of products. The further up the assembly chain this condition applies, the lower the cost for each product, as the cost of equipment peculiar to each product is minimized. Figure 11.1 (Chapter 11) shows the layout of a flexible assembly system (FAS), in which various subassemblies are kept separate until the configuration of the product is determined by the schedule.

The throw away product is particularly relevant to this principle. Correctly designed, these items are sealed only at the very last assembly operation. If parts are sealed before this stage, any fault caused or discovered is impossible to repair and thus defeats the object of the exercise and results in all the unsalvageable value added work being wasted.

FIFTEENTH DESIGN PRINCIPLE

The items and subassemblies must be handled without damage. It is all very well to see robots and grippers performing handling, presentation and assembly tasks, but unless components are specifically designed for automatic handling through protected surfaces and large smooth clamping surfaces, then scuffed, marred or otherwise damaged products could result.

CHAPTER 8
Workstations

A workstation is the place at which an assembly task takes place. It can be a single self-contained unit, a mixed man-robot unit, or one of many workstations that form an integrated flexible assembly system.

A self-contained workstation is effectively a worktable parallel to the ground and forming the lower limit of the robot's work envelope. The robot can be fitted directly to the worktable, mounted above it, or installed alongside. When fitted directly, the robot either 'sits' on top of the table (Figure 8.1), or (in the case of a cartesian configuration) is mounted along one edge (see Figure 5.11, Chapter 5). Whatever the arrangement, the robot's work envelope must cover the proposed range of activities.

Workstations and robots are, however, impotent unless the material to be transformed into 'value added goods' is presented in the right place at the desired rate. (Material feeders are dealt with in Chapter 9, and the economics of a workstation discussed in Chapter 15.)

The system

Raw material, partially processed assemblies and consumable materials are delivered to the workstation. After processing into value added goods, they are collected and transferred either to another workstation, or to the warehouse. Workstations contain all the auxiliary equipment (eg bowl feeders or orbital riveters) necessary for the robot to perform the assembly tasks. The activities of the workstation are monitored by the robot's controller, and the entire process is sequenced through the robot's input/output ports.

If the workstation is a mixed man-robot arrangement, then the safety and performance fluctuations of the operator have to be considered. Thus, even though the process is controlled

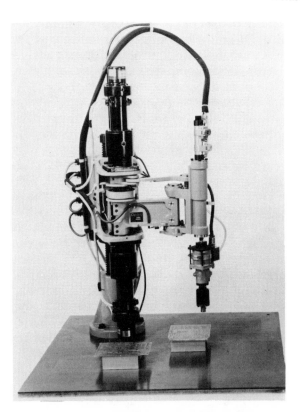

Figure 8.1 *A Pental HR 26 SCARA configuration robot has a payload of 15kg and a maximum reach of 400mm. It can be easily fitted to the surface of a worktable and controlled from a 'remote terminal'.* (courtesy of Sale Tilney Technology plc)

by the robot, each indexing cycle has to be authorized by the manual operative. Therefore, if the workstation includes a rotary table and the operator performs some of the assembly tasks at one of the stations, or just unloads and loads the jigs, the table must not be able to index until that person has authorized it. After the safety interlock has been released by the operator, the robot is free to index the table when the remainder of the automatic tasks for that cycle have been performed. After the table has indexed, the interlock becomes again the prerogative of the human operative.

An integrated assembly system contains a number of workstations (ie manual, mixed man-robot, or totally automatic). The exact mix is dependent upon the needs of the company. Such a complex system requires a sophisticated control system so that the activities of each workstation are monitored and controlled. Often, the entire system is controlled by a supervisory computer that in turn controls and downloads information to local workstation computers and to the computers controlling the robot at those workstations. In this way, the robots and other machines are independently controllable and can be operated as stand alone units. However, the overall scheduling and line balancing are performed by the master controller.

The flexibility of an integrated assembly system is given by both the software capability of the robots and by the design of the material transportation system that feeds it. The workstations can be linked by a rigid conveyor system that moves the material, components or subassemblies through the entire system; in which case, flexibility is given only by the robots. Alternatively, the material can be moved between the independent workstations by free roaming transportation devices that call to each workstation on (computer) demand.

There are a number of proprietary stand alone workstations used for manual, automatic or man-robot processing. One unusual version is the Lanco Hexagonal Transfer System (Figure 8.2), which can be extended to form an elongated hexagon, or be set up as a number of hexagonal nodes linked by transfer channels (Figure 8.3). The system uses platens that are 200mm square and fitted to a conveyor that transports them from one activity centre to another. Auxiliary equipment can be fitted either inside the hollow hexagon or around the outside. Naturally, the system is designed to interface with Lanco's own range of riveters, pick and place units and bowl feeders, etc, but other suppliers' equipment can be accommodated.

Assembly line balancing

It is important that the work content of each workstation within an assembly system is approximately the same, otherwise the line functions at the speed of the slowest station. If one assembly task takes far longer than others, then it should be replaced by a number of identical parallel workstations (see Figure 11.1, Chapter 11) so a bottleneck is avoided. The workstations in an integrated system can be either synchronous or asynchronous:

1. A synchronous system is one in which the assembly line functions as balanced, and indexes only when all activities at all workstations have finished for that cycle, thus progressing the assemblies through the

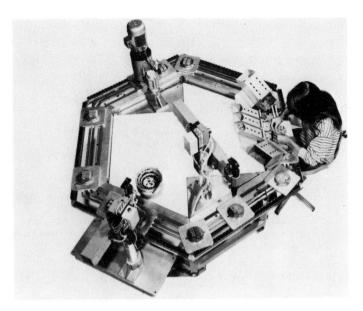

Figure 8.2 *A typical layout using the basic Lanco Hexagonal Transfer System. The operator takes components from gravity magazines, and the product progresses (clockwise) past two pick and place devices, one bowl feeder and one orbital riveter.*
(courtesy of Swissap, Royal Leamington Spa)

system, one station at a time, and at regular intervals.

2. An asynchronous system is one in which, whilst the workstations may be physically linked to each other, each station can function independently. This has an advantage in that any imbalance in the cell activities are accommodated by the cell buffers, which in turn allow production to continue downstream of a failed workstation until the buffers' stocks of the other workstations are used up.

Line balancing involves assigning the required assembly tasks to the individual workstations, which are sequentially ordered on a multistation assembly line. This assignment must be subject to the technological, precedence relationships between the work tasks. The major objective of line balancing is to minimize idle time, which is the total throughput time of the product *minus* the time spent on actual assembly processes.

111

Lanco Hexagonal Transfer
Assembly System 2000

Assembling an Air Valve

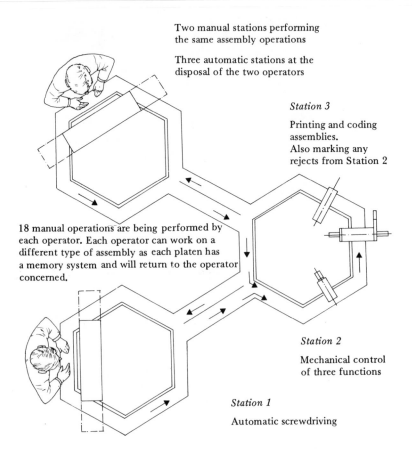

Two manual stations performing
the same assembly operations

Three automatic stations at the
disposal of the two operators

Station 3

Printing and coding
assemblies.
Also marking any
rejects from Station 2

18 manual operations are being performed by
each operator. Each operator can work on a
different type of assembly as each platen has
a memory system and will return to the operator
concerned.

Station 2

Mechanical control
of three functions

Station 1

Automatic screwdriving

Output:	About 30 assemblies per station per hour
Operators:	1 or 2
Number of platens: 20	

Figure 8.3 *A line diagram showing how the basic Lanco system
can be expanded. This arrangement has three automatic stations
and two identical manual stations.*
(courtesy of Swissap, Royal Leamington Spa)

112

IDLE TIME

In traditional manufacturing industry, idle time is analogous to the time components spend as 'work in progress' in the buffer stores. With a multistation assembly system, idle time is the time during which there is no activity at some, but not all, workstations. This idle time is created by one or two workstations having more work to do than others. Thus, progression is paced by the workstation with the longest activity.

It is essential to distinguish between apparent idle time and a robot's routined programmed pause during which some other activity is being conducted. For instance, a robot could be holding a component in an inspection device, making it appear idle. (This false impression also has man-robot safety implications, which are discussed in Chapter 12.)

Any reduction in idle time will yield positive cash bonuses to the company since, by minimizing idle time, the number of hours used to manufacture a batch of items is reduced. This also allows better utilization of the system, which means orders are processed more quickly. Obviously, at some point, it costs more to reduce idle time than would be recouped through increased productivity.

REDUCING WORKSTATIONS AND ACTIVITY TIME

There are two strategies to line balancing:

1. Reducing the number of workstations involved in the assembly task. To achieve this, the total assembly time is divided by the cycle time. The challenge is then to get as close to this calculated target value as possible, assuming the tasks can be simply divided by the number of workstations so computed, which rarely if ever occurs.
2. Reducing the amount of time spent at each workstation. This is limited by the speed and sophistication of the robots, auxiliary equipment and control systems used at each workstation. It also depends on the willingness and ability of the system designers to consider alternative assembly techniques, and other time-saving features often realized through lateral thinking or 'brainstorming'.

LINE BALANCING OF MANUAL ASSEMBLY LINES

The line balancing of robotic systems has been developed from techniques used with manual systems. Therefore, it is beneficial initially to examine the assumptions made for the line balancing of manual assembly lines. There are ten assumptions used in current balancing techniques:

1. The tasks, their performance times, precedence relationships, and other constraints are valid and accurate.
2. Task times are easily determined.
3. The precedence relationships cannot be violated.
4. The tasks are indivisible, ie breaking a designated task into its component elements will result in a longer task time.
5. Each station has one worker.
6. The work is performed at a standard pace by the workers.
7. There are no zoning constraints (ie work can be performed at any station, assuming the equipment is available).
8. Each station works on one item at a time.
9. The line produces a single product with no style variations.
10. Task times are independent of station assignment.

The first four assumptions concern the definition of the tasks. The assembly process must be broken down into discrete tasks, the descriptions of which must include and describe all the necessary movements of the assembler. The tasks must also be independent of each other, with respect to the precedence diagram.

Times are estimated by applying standard data to the motion requirements of the task, the data being derived from a work study analysis, or having been synthesized. These task times and the precedence relationships between the tasks are the primary input to the balancing method, the results of which depend upon the accuracy and integrity of the information provided.

Balancing a robot line

Manual lines do not need to be precisely balanced, since the natural versatility and adaptability of the human worker allow for deviation from the standard pace on the assembly line. A manual line can accommodate variation in either workstation balance or operator performance, but not so with a robot line.

This is because robots are very precise machines that perform tasks in the same way within each cycle at virtually zero time variance.

If the solution to a robot line balancing problem gives a particular robot 3% idle time, then that robot will actually be idle for 3% of the cycle time of that system. However, whilst a robot performs a task in a constant never varying time, the robot can be adjusted so that the cycle time is lengthened or shortened, even though the 'new cycle' time becomes the constant. This ability to vary the operational speeds of robots is useful where a robotic assembly system incorporates hard automation and/or manual workers. In this case, the operational speeds of individual robots along the line are adjusted to compensate for the variations in the hard automation so that the idle time incurred at the manual stations is minimized. In this manner, the flexibility of the robot duplicates the versatility of the manual worker, without incurring the penalties of cycle time variance.

CONTROLLING SPEED

The speed at which the tasks are performed at each station is no longer left to the ingenuity of the individual worker. This is now a centralized decision variable that is considered within the total design philosophy of that line. Thus, production costs and delivery schedules can be forecast and 'guaranteed'.

An assembly line manufactures items in real-time and, therefore, it makes sense to monitor and control items in real-time. However, there is one problem, namely that real-time control requires a number of simultaneous decisions to be made by the line controller, and that the more complex the system the larger the number of simultaneous answers needed. Further, and irrespective of whether the controller is human or microelectronic, the greater the number of simultaneous decisions that have to be made, the greater the risk of a functional overload.

Thus, the control logic of the system should be such that each local problem and variance is solved by the local computer, whilst the general running and balancing of the system is maintained by the supervisory computer. This division of computer responsibility for robot lines is not uncommon. For example, the variation in presentation rate and progression speed of components across the work zone in a surface coating line, due to the fluctuations in conveyor speed, does not affect the results of the robot task. The reason is that the robot's program is synchronized with the speed of the conveyor so, as the conveyor speeds up or slows down, the 'reading rate' of the robot's program is automatically varied.

Implementation of workstations
The introduction of a robot workstation into an existing line can cause disruption to output and require changes to the upstream and downstream elements of that line.

UPSTREAM ELEMENTS

The upstream elements of a line perform a number of tasks on the product being assembled before it reaches the robot. The tasks to be performed at the robot are only successful if the quality and tolerances of the material arriving at that workstation are compatible with the needs of the robot. Obtaining compatibility may mean changing the level of quality, increasing the degree of consistency, changing tolerances, or adding new features to the components which may mean tool changing at the upstream workstations.

DOWNSTREAM ELEMENTS

The downstream workstations are affected by the output rate of the robot. As is often the case, traditional assembly processes are the slowest activity on the line. If the robot increases the (shift) throughput rate, the number of subsequent workstations may have to be increased to remove any out of balance. As with the upstream workstations, any changes in component design may affect the jigging and tooling.

If a robot is used instead of manual assemblers, then one element of disorder has been removed from the process. This should mean that process quality has a higher consistency, and it is likely that any inspection stations downstream of the robotic station could be reduced or removed. Introducing a robotic assembly system does not necessarily negate the need for inspection. Rather, the use of robots requires a more consistent quality of components, and that many of the robot workstations should contain integral inspection routines. It should never be forgotten that a wrongly programmed robot line will consistently and continuously produce scrap items, unless the line is monitored.

CHAPTER 9

Material feeders

As discussed in Chapter 8, assembly by robot can only be successful if the material and components used in the product are presented to the robot in the correct attitude and orientation, at the right place and time. Products consist of three material groups:

1. Discrete items such as screws and transformers.
2. Bulk materials such as sealant and adhesives.
3. Preprocessed material such as subassemblies.

A successful robot installation assumes that everything is in its correct place, and that the process will perform according to the written specification. Unfortunately, in the real world, situations are not always black and white. Consequently, Chapter 10 looks at the use of sensing techniques that will enable a robot to perform assembly tasks in an unclear environment.

The presentation of discrete items is achieved with automatic feeders.

Automatic feeders
Automatic feeders transform a mass of items into an ordered presentation of individual items that can be collected by the robot.

Three general examples of feeders are bowl feeders, autoscrewdrivers, and gravity feeders.

BOWL FEEDERS
The vibratory bowl feeder (Figure 9.1) uses its vibratory action to move items from the bottom of the bowl up an internal ramp leading to the top. As the components move up the ramp, they pass through a series of passive devices which, using gravity and guide rails, encourage the components to adopt the correct

attitude and orientation. At various points along the ramp, passive 'gates' are fitted which are designed to pass only correctly orientated items for that gate, ejecting others back to the bottom of the bowl. A great deal of expertise and unquantifiable knowledge is needed in the design of the track, correction devices and gates.

Figure 9.1 *A vibratory bowl feeder that supplies a workstation with two rows of components. At the back centre and back left of the bowl can be seen two sets of passive devices that eject incorrectly positioned items.*
(courtesy of Aylesbury Automation Ltd, Aylesbury)

The processing rate of all bowl feeders must be in excess of the required take off rate by the robot, so as to compensate for the number of incorrectly positioned items that are returned, by the various selection devices and gates, to the bottom of the bowl. If the 'success' rate of the feeder is higher than the take off rate, the component backup activates sensors that stop the operation of the bowl until the surplus has been reduced, at which time another sensor is activated and the bowl switched on again. To ensure that a component is present prior to the robot moving to the pickup point, a proximity switch at the pickup station is 'read' by the robot's control before the pickup move is made. If no component is present, the robot often enters an abort routine in its program, instead of 'waiting' for a component to be presented.

Bowl feeders are used throughout the manufacturing industry and are one of the more successful methods of presenting discrete items to a robot. As a general statement, they are designed for one component and usually require that the component has very precise tolerances on the surfaces that are to be used for guidance by the bowl feeder. This is particularly so if the component has to be presented in a very precise orientation, since a wide tolerance on the guiding surfaces could allow the component to swing to an undesirable orientation. If, however, the exact orientation is not important and/or some other device is used to achieve precise orientation prior to the assembly process being conducted, then the feeders can be used for a variety of similar parts.

AUTOSCREWDRIVERS

Autoscrewdrivers are used for feeding discrete items such as screws, rivets, nuts and studs (Figure 6.4, Chapter 6). These devices use mechanical selectors to present individual items to an air/vacuum system that sucks or blows each item to the driving end of the device, for example, a screw is presented to jaws that surround the screwdriver head. The axial motion of the autoscrewdriver inserts the screw into its hole, then a simultaneous retraction of the autoscrewdriver withdraws the jaws from the inserted screw and causes the 'next' screw to be delivered to the jaws.

The success of this method depends upon items not being able to 'tumble' as they are fed to the driving head, and also on the items being easily orientated prior to use.

The combination of a robot and an automatic screwdriver is already an established procedure for assembly tasks. If the robot cannot withstand the axial force necessary to ensure that screws are inserted to the correct depth and torque, the robot is used to position accurately an auxiliary device that holds the autoscrewdriver. This device is specially designed to apply an axial force to the screws so that the desired insertion is assured. In this manner, the assembly robot uses its attributes of precision, speed and control, but its application is not compromised by the need for high rigidity.

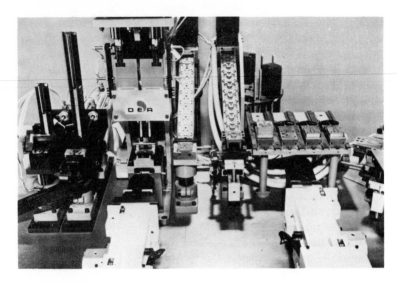

Figure 9.2 *A compressor valve assembly cell that uses two robots (foreground), two gravity feeders (lefthand), one bowl feeder feeding two rivets (centre), two gravity chutes fed by pick and place devices from magazines at the rear (right of logo), a table with two different components in two piles, and a four way outlet for screws from a bowl feeder (righthand).*
(courtesy of Fairey Automation Ltd)

GRAVITY FEEDERS

Some discrete items cannot be bowl fed because they are an awkward shape or the space within the workstation will not allow the feeders to be fitted. If this is the case, gravity magazines can be used. These devices are usually vertical 'tubes', magazines or rails containing a number of items arranged in the correct attitude and orientation. The 'bottom' item from the column is pushed away and towards a presentation position from where it can be collected by the robot. Alternatively, the robot can collect each item from the gravity feeder without the need for the auxiliary presentation mechanism.

Gravity feeders are simple devices that work extremely well provided they have been correctly designed, and properly filled prior to use. There are two types of gravity feeders:

1. Where the magazine is an external shell around the components.
2. Where components are fitted onto a rod.

Components invariably have one dimension smaller than the others, and they are stacked parallel to this smaller dimension. Ejection is usually achieved by a blade pushing the lowest component from beneath the pile. The magazines need to be designed so that they cannot be incorrectly loaded. However, in many cases, it is still left to the manual loader to check the correct attitude and orientation of the items before putting them into the storage device. The capacity of the magazines is limited and as they cannot be easily reloaded automatically, the system needs interlocks in order that empty magazines can be replaced by full ones, as required.

The variation and complexity of these material feeders at a robot workstation is well illustrated in Figure 9.2. At this workstation, two robots working in tandem assemble a compressor valve. A valve plate is collected from the first chute, and transferred to the press. The same robot then collects an exhaust valve from the lefthand gravity feeder, which is then placed over the newly positioned valve plate. Next, a leaf lock is transferred from the second lefthand gravity feeder and placed onto the other two items in the press. The press is then activated, resulting in three items being riveted together.

Simultaneous with these activities, the second robot collects a gasket from one of the two lefthand stacks on the table and places it on top of the valve body situated at the end of the second chute, which is to the immediate left of the table in Figure 9.2. The riveted subassembly is then collected, inverted and placed onto the gasket-valve body subassembly. Finally, a suction cup leaf is collected from one of the two righthand stacks on the table and placed on top of the now complete assembly sandwich.

The entire four element assembly is transferred by the second robot to the autoscrewdriver station, at which four screws are issued by the bowl feeder and used to clamp the entire compressor valve together. When the assembly is complete, it is removed from the autoscrewdriver cell by the same robot and transferred to an exit conveyor. The compressor valve is shown 'exploded' in Figure 9.3(a) and assembled in Figure 9.3(b).

Feeding delicate items

Vibratory bowl feeders and gravity feeders have one fault in common, ie the components can bump into one another, and/ or are moved relatively across or along fixed frictional sur-

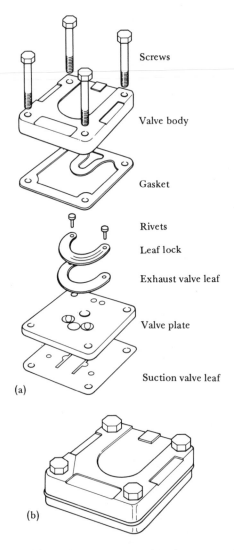

Figure 9.3 *(a) Components used in the construction of the compressor valve; (b) the assembled unit.* (from a DEA PRAGMA catalogue)

faces. In extreme cases, this can cause functional damage or a cosmetic surface to be marred or scuffed which, if not noticed, could result in the waste of all the added value work.

There are three alternative methods by which delicate components and materials can be presented to the robot without risk of damage:

1. The use of a conveyor that contains a number of specially scalloped containers that hold an individual item in the desired attitude and orientation. The items are manually loaded into the scallops and are extracted by the robot at the workstation.
2. The provision of a kit of parts in which each item has a unique place and attitude and orientation. The robot can pick each item from the kit and proceed to perform the assembly tasks for that workstation.
3. The manual loading of the 'delicate' item at the assembly station into a jig which is subsequently presented to the robot.

Component manufacture at site of usage

There are a number of categories of discrete components that are difficult to feed by any method. One example is open ended springs which are difficult to feed since they can tangle into masses and not easily be separated automatically. One way to overcome this problem, assuming there is no design solution (as suggested in Chapter 7), is to make the components at the place and time at which they are used. Another reason for manufacture at site is to prevent contamination of, or damage to, items being processed. A further example is where a number of variants are made from a common element (eg a cell is provided with a bulk supply of identical parts that are processed *in situ* to one of a series of slightly different items peculiar to the product being processed).

Manufacture at point of usage means the appropriate device has to be fitted to the workstation which, when activated by the robot, will ensure a component is made there and then. The component may then be collected by the robot and fitted to the other parts presented by more conventional methods. However, this method of manufacture is limited to small items and/or those that have a short process time. This method often takes the form of a bandolier of partially blanked or discrete items fed into a tool. The items are then processed and/or sheared off the bandolier, and directly inserted into the product being assembled.

Manufacture at site has a number of economic advantages:

1. Complex feeding systems (for discrete parts) are

avoided and replaced by a relatively less expensive feeder of bandoliers, or raw material.

2. Storage space is needed only for the base material, which can often be used for other totally different items or products.

3. The raw material does not have value added to it, nor is it committed to any one product until required.

Feeding consumable materials

Almost every product uses some form of consumable material (eg arc welding rods, sealant, adhesives, or paint). If a product is manually assembled, then consumable items are applied from containers sized to suit the convenience of the assembler. In the case of adhesives, small capacity tubes are used which have two economic disadvantages, ie the amount of material applied is usually much more than is needed to satisfy the technical requirements, and the containers are often thrown away before they are completely empty.

A similar comparison to the use of adhesives may be made by examining the use of arc welding rods, which are again used in the manner desired by the welder rather than that necessary to satisfy the technical needs of the task. The rods are also often thrown away before they are fully consumed, resulting in a certain amount of wastage.

If a robot is used to perform the above tasks, both of which are assembly related, the amount of material applied is that which is programmed in, and the 'bit for luck' does not get applied, as is the case with manual assembly. Therefore, material usage is known and constant. The robot can use bulk containers of sealant, or continuous reels of welding wire, which will be more economical from a purchasing point of view. In addition, there would be smaller losses from the 'nearly empty' drums of sealant, or reels of wire, that are discarded, because the total waste represents a smaller percentage of the material cost.

Consumable bulk materials (eg adhesives and sealants) have their own set of feeding problems because the materials are cured by either proximity to air, or by contact with metallic objects. Alternatively, the materials are not 'solid' or discrete, so there are difficulties with presenting the correct amount of material in the right place.

Technically, these problems are solved by simply fitting a pump to the sealant or adhesive container. Air curing is minimized by reducing the amount of material exposed to the air prior to dispensing. Since the nozzle is always in contact with the air, many devices incorporate purge routines that eject a small amount of material before application begins. The risk of unwanted curing due to contact with metal is minimized by using nonmetallic liners (eg nylon or teflon) throughout the contact surfaces in the material containers, pump, pipes and nozzle.

The pipes and nozzles of these pumping systems are easily fitted as end effectors of the robots. The robot's precision is therefore used to ensure accurate placement of bead paths, or spots of material, whilst the robot control system monitors and controls the speed of dispensing and opens or shuts off the nozzle.

Some adhesives and sealants need two or more materials to be mixed immediately prior to dispensing. Again, these mixing and dispensing systems are easily fitted to robots, which will give precision and control to the application of a 'fluid' material.

Conveyors

Conveyors can index components to a workstation, or have continuous motion through the workstation. In the former case, a component is presented at a known rate and remains within the envelope of the robot for a fixed time, during which it can be offloaded from the conveyor, or be processed whilst still on the conveyor.

Continuous motion conveyors present the components at a known flow rate, but the robot must track the item as it passes through its work envelope, whilst either unloading or processing that item.

A composite arrangement is where a continuous conveyor is used and the components are halted at the workstation by a barrier across the belt. The unloading and processing is then conducted and, if required, the barrier is removed and the item continues along the conveyor. This last option allows for buffer stores to build up at each station and take some of the rigidness out of a complicated assembly system.

In all cases, so the robot knows what to do with the presented items, the system controller and/or local computer must be aware of the identification and sequence of the items on the conveyor. This is done by assuming that all the items are identical and no problems could occur. Alternatively, the sequence can be put manually into the scheduling computer, or each item can have some identification that can be read automatically as it enters the work envelope of the robot. Of the three alternatives, it is the last one that makes the most sense as it offers the most flexibility to the loading of the system and minimizes the risk of preoperational scheduling error. Identification can be by bar codes read by a laser scanner, fixed binary keys that activate microswitches, or EPROM memory boards that contain all the necessary information, and which can be updated as the item (and its pallet) passes through each station.

Conveyors have several disadvantages, mainly because they are in fixed routes and not easily changed. They also tend to block access to the workstations.

Automated guided vehicles

The automated guided vehicle (AGV) is a flexible and virtually invisible alternative to conveyors, as far as congestion is concerned. The AGV is a free ranging vehicle that follows magnetic tracks laid in the floor (Figure 9.4), and which moves material around a factory. It is not limited to one route, but can be automatically switched throughout an entire system of interconnected routes. Thus, a small number of AGVs can service a large number of workstations.

An AGV can move material through an assembly cell at each station, where it will wait until the task is finished before transporting the value added material to another workstation or back to the warehouse. Alternatively, an AGV can be used to take material to workstations where it is offloaded by a simple pick and place robot or, by matching the levels of the AGV platform and the spur conveyor that services that workstation, the material can be transferred by a simple pull or push mechanism. After processing, the material can be collected by another AGV and transferred to another workstation, or taken to the stores.

The use of automated guided vehicles is not limited to the factory floor or warehouse (see Figure 9.5) as they can move along carpeted walkways of the commercial or managerial sectors of a company, assuming these walkways form part of a logical route for materials being processed. The AGVs can still 'read' their route through the carpet, as they move slowly towards their destination. The vehicles are fitted with collision bumpers that either gently move an object out of the way, or cause the AGV to stop.

Figure 9.4 *An automated guided vehicle in a warehouse. Magnetic tracks can be seen on the floor. The AGV fits between the wheels of a standard pallet truck and pulls/pushes the pallet truck by a vertically extended rod.*
(courtesy of Telelift GmbH & Co, Puchheim, West Germany)

Prepackaged material control

Historically, raw material and components have been delivered from the vendors in a random orientation and attitude. Therefore, before a process can start, each item has to be reconfigured so it can be correctly presented. Further, as is often the case, the processed items are then stored in a disordered fashion and transported to their next destination, where they again need to be put in order before the next process can start. Kitting or contoured material transportation trays are ways of ensuring that the ordered status remains throughout the processing of the items.

KITTING

Kitting is the presentation to each workstation of the exact number of items needed to process a single product. For a robotic assembly task, each item must be arranged in a unique place on the kitting tray so that the robot can unload the tray without having to search and identify each part.

Figure 9.5 *An AGV moves unobtrusively along a carpeted walkway, from the warehouse to a workstation. The relationship of the AGV to the pallet trolley is shown quite clearly.*
(courtesy of Telelift GmbH & Co, Puchheim, West Germany)

TRANSPORTATION TRAY

If the items to be processed are delivered to the robot workstation on a tray, it makes sense to use that same tray to move the items after processing to the next activity. The contoured tray is designed so that a number of build stages of the product can be held and transported on one tray. Simplistically, it comes down to identifying the various features at each build stage that can be used to locate the items/subassembly in the tray. If the tray remains on the transportation system, and is not disturbed, control of material through the entire assembly process can be obtained.

This ideal of material control can be extended back to the material vendors, with the purchaser specifying that each item be supplied in a contoured tray compatible with the internal transportation system of the robot assembly line. Obviously, the cost of the items, as delivered, will be more expensive, but the savings gained through not having to orientate the components within the assembly process will cancel out this penalty.

Sensing and vision

The majority of robots are machines without any *built in* sensory capacity. They are instructed what to do and when to do it, and their programs assume that all parts are correct, in their proper places, and presented in the correct sequence at the right time. However, this does not mean that a robot cannot be used efficiently as a productive element within the manufacturing industry, since many robots have been installed without sensors fitted, yet they perform their tasks effectively.

The performance of assembly tasks, especially those requiring precise insertion of one item into another, can benefit from the use of sensors. For example:

1. Prevention of damage through incorrect assembly.
2. Cost savings through the automatic identification of randomly presented items.
3. Minimization of the cycle time through the availability of real-time information of the relative positions of the parts to be mated.
4. An increased system reliability through the removal of mechanically complex feeding and sorting devices.

This chapter highlights and discusses sensors which are available at present and which are used with assembly robots. The more sensing capability a robot has, the more 'intelligent' it becomes. If a robot can accurately evaluate its working environment and react accordingly, dynamic situations that accurately reflect the true world of the manufacturing industry can be accommodated.

Sensors
The idealized gripper shown in Figure 5.8 is fitted with a multitude of sensory aids. Of course, in reality, this gripper would

be expensive and have technical implications, but the intent is valid. Consider, for instance, two components of similar shape, but with different dimensions, being processed by a single robot. This gripper could be used to identify a particular component by the size indicated through a diode array or simple tactile sensors, whereas weight could be determined by load cells within the gripper.

A robot needs to be aware of two aspects of its environment if it is to perform assembly tasks. Firstly, it must be aware of what is happening within the general workstation (eg what other activities are occurring and where the other robots' arms are) and secondly, it must have a knowledge of what is happening when a task is being performed (eg the robot moves a gripper or a tool to a point within the workstation and then performs a task). Unless the robot can sense what is happening, the process is being conducted in ignorance.

Knowledge of the general workstation is achieved through sensory devices, external to the robot. All sensors are 'read' by the robot's input signal port, prior to it making each action. Hence the degree of risk to the system is a function of the sensory network used and the program. Monitoring the task is realized both through external sensors and those that are integral within the gripper or tool.

External sensors monitor the motion of the robot, and other equipment within the workplace, to prevent collision. Optical sensors can be fitted directly to the gripper (through the use of fibre optics) so that the orientation of the gripper, relative to a randomly presented object, can be detected and changed so that pickup can occur. If an assembly task is to be performed within a 'box', then external sensors might become occluded by both the profile of the box and the gripper, hence explicit control can only be maintained by a sensor forming an integral part of the gripper.

The need for sensors to be fitted to robots used in assembly tasks has been realized by at least two robot manufacturers, who market these units equipped with integral sensors. The PRAGMA robot, manufactured by DEA in Italy (sold by Fairey Automation Ltd in the UK) and General Electric in the USA, has parts detection sensors fitted to its interchangeable gripper module, and force sensors fitted to its wrist.

IBM's 7565 gantry robot is also fitted with an intelligent gripper module, which has parallel motion and includes both tactile and optical sensors. The tactile sensors are fitted to both halves of the gripper module and crudely duplicate the function of the human palm and fingertip in that the open gripper can be used to detect objects through contact with the internal surface of either gripper half. Objects can be 'pinched' through using sensors at the internal tips of the gripper, or the gripper can hold and manoeuvre an object in a controlled manner, with the clamping force being monitored by all the tactile sensors. Part sensing is achieved through a light beam between the two halves of the gripper module. A break in the beam indicates that an object is between the gripper jaws, and the tactile sensors indicate its state.

Sensors can be roughly classified as follows:

1. Tactile.
2. Proximity.
3. Pneumatic.
4. Optical.
5. Optical sensors and fibre optics.
6. Vision sensors.

TACTILE SENSORS
The most common sensors used by robots are tactile units that determine if an item is present or not. The simplest version is the microswitch, which makes or breaks a circuit when the switch is operated by a component arriving at the sensing station. Another version uses a matrix of wires, covered by a plastic skin, that approximates to the human 'hand' and determines not only presence, but also position. Provided the information about the item is in the robot's memory, the item is identified.

The force sensor shown in Figure 10.1 has a microcomputer at every node of the matrix. It uses a custom designed, very large scale integration (VLSI) device to perform transduction, tactile image processing and communication. Forces are transduced using a conductive plastic technique in conjunction with metal electrodes on the surface of the integrated circuit.

Tactile sensors are often used to indicate what happens within the gripper as it operates. For instance, the sensor can be used to inform the robot that the gripper has closed on an object, or that it has closed too far and therefore (by deduction) the object was either not there or has been crushed, or has fallen out of the gripper.

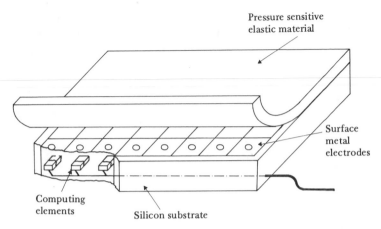

Figure 10.1 *The architecture of the VLSI tactile sensor is revealed. A sheet of conducive plastic is placed in surface-to-surface contact with a custom designed VLSI circuit. Each element within the array has its own computing element, hence this sensor can be used to detect both presence and form.*
(from Railbert and Tanner 1982)

Tactile sensors are two state units: either on or off. Thus, they cannot determine if a component is approaching, or being approached. The ability to determine the presence of objects at a range is the reason why proximity switches are popular in dynamic situations.

PROXIMITY SENSORS

Proximity switches are noncontact sensors that work on the principle of either capacitance or inductance. They are activated when an object comes within range of the sensor. Inductive sensors detect the presence of metals, whereas capacitance sensors can be used for both nonconducting materials (eg wood and glass) as well as for ferrous and nonferrous metals. The operational range of these sensors varies from two to several hundred millimetres. They can therefore be used as an advanced warning system by the robot to 'inform' it that a component is approaching, or to detect the arrival of the same object at its destination.

PNEUMATIC SENSORS

Airjet sensors detect presence by breaking or disturbing the air flow. They are used with bowl feeders to monitor and control the buffer store of parts to be collected by the robot. When the air steam is changed, the feeder is switched off until another sensor (further up the track) reactivates it as the buffer stock reduces.

OPTICAL SENSORS

Another type of noncontact sensor is the optical sensor which detects presence through the breaking of a modulated infrared beam between the emitter and the sensor. Alternatively, objects can be detected by retroreflective sensors, through reflectance of the emitted modulated infrared beam by the approaching object back to the sensor. In both these cases, either presence or nearness is detected.

It is important to understand the difference between these simple optical sensors that use 'light' and the rapidly developing vision technology discussed on page 135. Briefly, optical sensors have limited applications, whereas vision equipment may be used for scene analysis.

OPTICAL SENSORS AND FIBRE OPTICS

The use of fibre optics means that the sensor beam can be placed in a position that allows optimal monitoring of an activity. Fibre optics sensors are essentially used to control and monitor a completed assembly task. (Fibre optics within an intelligent gripper are discussed in Chapter 5.)

A Hirata AR-H250-4 SCARA type robot (Figure 10.2) can be used to assemble a range of electrical connectors. The connectors are of two base lengths, fitted with either one or two rows of contacts, which can be either solder lugs or printed circuit board pins. Two staking pins are also inserted into the connector which is polarized by a notch at one end.

The successful assembly of these items by robots has been simplified by using a number of optical sensors, linked by fibre optics to the robot's control system. The assembly cell consists of one robot, three bowl feeders and one hard automation device. Since the hard automation device cannot be software reconfigured for the different combinations of connectors, the products are processed in batches.

One bowl feeder is used to present both types of base, which are supplied right way up, with either end leading. Another bowl feeder presents the contacts. Again, one unit is used for the two types of base, which are offered in the correct attitude and orientation. The remaining bowl feeder is used for the staking pins.

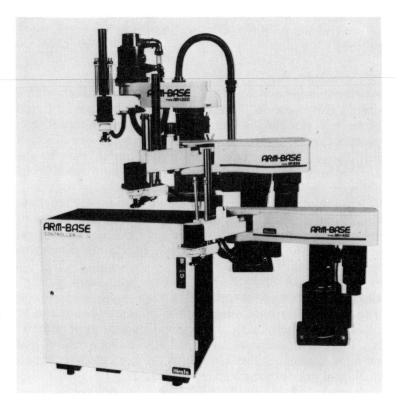

Figure 10.2 *Hirata assembly robots. From back to front:*
the AR-H250, AR-H300 and AR-H450.
(courtesy of Geoffrey Champion)

The robot's program has been constructed to maximize the robot's flexibility so, in the future, it can be used to process an additional, or extended, range of connectors. Each assembly task is allocated its own program, hence the 'program' for a given connector is a small 'master' program containing a lot of jump statements to subroutines that describe the appropriate tasks.

In operation, the robot is 'informed' of the type of connector to be processed. One base is collected by the robot from the pickup point on the bowl feeder and placed into the assembly fixture. Four sensors are fitted to this fixture to identify both the length of the base and its orientation. One sensor is activated by the 'front end' of the base, which confirms to the system that an object is present and situated at the gauge face of the jig. A second sensor, adjacent to the first, uses the polarization notch as a check for the orientation of the base. If the notch is identified, the insertion program will be performed in its default mode. If the notch is not found, this insertion program is rotated 180 arc degrees (in software) so that the insertion procedure is correct for that orientation.

The third and fourth sensors are used to determine the length of the base. At each end of the base (symmetric about its width and length) is a hole, which is searched for by the two sensors. A hole detected by one sensor allows the program to determine which length base has been picked up by the robot. If the base is deemed to be the wrong type, it is rejected. Similarly, if neither sensor is activated, the system assumes that the item is faulty and rejects it. The third cause for rejection is if both sensors are activated, which indicates that the base is too long.

Assuming that the correct base has been placed in the assembly fixture, the robot then collects a contact from the bowl feeder and inserts it. This process is continued until the programmed number of contacts have been inserted. Next, the staking pins are collected from their bowl feeder and inserted into their respective holes.

After insertion is complete, and assuming that solder tag contacts have been used, the robot moves the base – with its contacts and staking pins – to the hard automation twist fixture, which locks the contacts into the base. The connector is then removed from the twist fixture and placed on an exit conveyor.

This system was installed in early 1984 at a cost of £35 000. Since then it has been operated 20 hours a day, producing between 14 000 and 15 000 assemblies a week at an insertion rate of one contact every 2.4 seconds.

VISION SENSORS
Of all the available machine senses, it is perhaps vision that generates more information per unit of time. As already mentioned, it is important to distinguish between simple optical sensors and the rapidly developing science and industry of vision technology. The simple 'here/not here' optical sensors discussed earlier have limited applications. However, since the end of the 1970s vision equipment has been used within an industrial environment for scene analysis.

A robot can be used to process a wide variety of different products; for example, products can be processed in a batch mode, after which the system is reconfigured for the next batch. Alternatively, the batch can be processed as a randomly presented queue of work. If objects are presented in a random sequence, the robot needs to know what object is being presented so it can perform the correct process and activity for that object. In other words, whilst two consecutive objects might need painting, they could require different colours as well as different coverage patterns.

Provided that the robot has been 'taught' to recognize the product, it will process the item correctly. However, recognition by a robot is achieved by matching a two dimensional digital image in the robot's memory with a two dimensional digital image presented by the approaching object. Unless this object is presented in a very specific orientation and attitude, the images will not match and the process will stop or the part will not be processed because the robot does not recognize it.

Simplistically, a vision system is a number of photosensitive elements (photodiodes) arranged in a matrix, either in a single line or in a square or rectangular manner. These photodiodes are 'measured' by the number of pixels that describe their shape. For instance, a single array could be 1×20, or 1×2000. Alternatively, a two dimensional array could be 380×380, or 1024×512.

A vision system reacts to light rays reflected from an object or blocked by its shadow, or emitted by it. For instance, if an object is put on a translucent conveyor belt, its shape will be detected against the backlit surroundings. Conversely, if the object is illuminated, then it is distinguished by the dark background. Alternatively, if the object is a different temperature to that of the immediate background, it will emit infrared rays that can be detected by various detectors sensitive to nonvisual spectra. For a single line array, there must be some relative motion between the array and the object, so that each line scan made by the array can build up in the memory to form a composite whole of the object to be identified. Either way, identification is made by comparison with a digital pattern formed by the activated elements, with the patterns held in memory. A match above a predetermined confidence level will identify the object and inform the robot what action to take.

There are three reasons why a match is not made:

1. The object is new and its image is not in the vision system's memory.
2. The object is presented in a distorted manner (eg resting on top of another item).
3. The object is presented in the correct plane, but with wrong orientation.

If a match cannot be made, then as long as there is sufficient time within the process, the image of the approaching object can be manipulated by the vision system to determine if a match can be made. Alternatively, the object can be placed on an auxiliary piece of equipment and rotated beneath the vision system, again to determine if a match can be made.

The success of identification depends upon the process knowledge held by the designers of the vision system since if, for instance, a plate is presented to a robot by an overhead conveyor, its silhouette will vary depending on the amount of freedom it has to rotate in the vertical plane. If it is assumed that mechanical guides cannot be used to minimize this motion, the vision system would have to incorporate two geometric patterns for each different object. These patterns would represent the object's silhouette at the extreme points of its rotation and, in the case of a cylindrical object, would vary from a complete circle to an ellipse or, with a prismatic object, from a square to a rectangle.

Much research into various vision systems has been done (and continues to be done) in research laboratories worldwide. Perhaps the biggest 'red herring' of all the vision related research is the solution to the 'bin picking' problem. This revolves around the apparent need to unload discrete components from a container in which they are stacked in random positions and orientations. The vision system has to evaluate the scene and distinguish one component from another, and then control the robot so the required item is gathered from the bin. Whilst the academic value of such research is appreciated, its direct industrial economic application to industry is hard to justify, especially as there are a number of simpler methods of resolving the problem.

Selecting a suitable sensor

With so many sensors to choose from, the choice depends as much upon the object being processed as it does upon the property required to be sensed. For example, consider the detection of a very lightweight component at some pickup station. It might be blown away by an air stream. Alternatively, the component may be of insufficient weight to close the microswitch, in which case the component will jam between the feeder wall and the switch lever and only release when the axial pressure from the following components forces it past the switch. The solution here might be to use an optoelectric sensor, or change the feeder design.

Most sensors are easily interfaced with the robot, through its input/output ports. Many of these ports have a byte of information that indicates the status of one or several sensors.

Depending upon the match between the sensor status read by the robot, and that stored in its memory for that program step, the robot will either perform the appropriate task, wait for the sensor status to change, or warn the supervisor that something is wrong.

Automatic inspection

Sensors are also used to determine if an assembly operation has been completed satisfactorily, which gives both economic and technical advantages to a robotic assembly system. An assembly task usually requires that a robot transforms an inert heap of raw material and components into a value added product, through correct processing and assembly of those items. Yet, the revenue gained from that product is dependent upon its functional reliability, which in turn depends upon the quality of material and components used, as well as the reliability of the assembly process.

An assembly task needs to be inspected at two levels:

1. Task completeness, which means checking whether or not the activity takes place. For example, were the screws fitted? Was the bearing installed? Were the feeders empty and the robot merely followed its program without anything being assembled?
2. Inspection to check assembly functionally at each build stage. For instance, the application of hot melt adhesive near to an electronic component could cause functional damage to that component. If, after the hot melt disposition, the component is checked electronically, any functional damage should be immediately apparent. The problem can then be

remedied at that time, before any more value is added to the product. If, in this example, the functional check is not performed, product failure at a later stage will be more expensive to remedy since the product might have to be totally refurbished, or a particular subassembly replaced. In either case, inspection and remedy at an earlier stage of inspection are less expensive in terms of material and manpower.

ONLINE INSPECTION

Online inspection at the function and achievement levels is possible in real-time through the use of sensors, provided that the product is correctly designed and the robot has the necessary interface capabilities.

The achievement level of any assembly task could be determined by tactile sensors which ensure that a preset dimension of 'insertion' has been realized, or a load cell can indicate whether an axial force or insertion torque has been reached by the assembly tool. Many assembly tasks have distinct audio signatures that indicate if assembly has been correctly achieved. An audio sensor could therefore be used to indicate completeness of an assembly task. For instance, plastic mouldings snap together or deflect, depending on the alignment of the parts being assembled and forces applied. A correctly assembled part requires a certain axial motion of the gripper as the connecting elements enter each other. If the part is misaligned, the item will either resist the closing force of the robot or deflect under that same force. At some point the robot will, through deflection, reach its preset extension and assume that assembly is achieved. Alternatively, two components crush together or one part is pierced by the gripper. In all cases, a combination of an audio sensor (which detects the snap signature) and a preset gripper extension provides feedback to indicate correct assembly (or not).

Many tests require access to a component while it is being processed in the fixtures within the workstation. Provided this need has been identified in the early product design stages, there are no technical problems in performing computerized online inspection in real-time. Microelectronics allow the data to be analysed, collated and presented in different formats that allow different interest levels of management to assess the productivity of that process. If there is a problem, it is normally a random fault occurring on only one item, which the process can ignore. However, if a fault occurs on several items, the process has to be shut down while the problem is identified and resolved.

Many problems occur through drift of in-process parameters. Again, the continuous monitoring of the process allows the drift to be spotted long before it becomes disastrous and be resolved at a convenient time (eg at the end of a shift).

In spite of the availability and sophistication of today's sensors, the application of robots to assembly tasks ultimately depends upon the economic benefits derived from the installation. The identification and quantification of these benefits and their method of presentation as a financial proposal are discussed in Chapters 13-15.

The sensory capability of even the most intelligent robot currently available is limited. Therefore there is a continued need for humans to interact as productive elements alongside robots. The inherent problems at man-machine interfaces are dealt with in Chapter 11.

CHAPTER 11
Man-machine mix

The use of robots to perform assembly tasks does not mean that humans are no longer needed as participants within the process. The totally computerized and automated design, manufacture and assembly of products without any human input (except as consumers) exist only in the minds of science fiction writers or predictors of the far future. Undoubtedly this may eventually happen, but until then the designers of robotic assembly systems must acknowledge and take into account the necessary interaction of people and robots.

In any man-machine system, safety is an important issue. Two aspects of safety, namely the protection of humans from robots and the prevention of damage to the robots by humans, are dealt with in Chapter 12.

Interaction between man and machine
Interaction can take place on several fronts within the processes associated with assembly. It is important that these interfaces are recognized and that robot systems are designed to maximize the advantages humans can offer to an assembly system. At the same time, the natural and psychological disadvantages of using humans must be minimized.

Today's assembly robots are technically sophisticated. Many assembly systems utilize a robot's ability to perform a desired technical task, at a given quality level, and with a predictable cycle time. However, for various reasons, the components and products being processed may not always be suitable for automatic loading and unloading at the workstation, or for subsequent automatic transfer to other work and storage areas. For example, there may be technical problems associated with the fragility of the items being handled, and it may not be economically feasible to redesign them.

One robot task that illustrates the component handling problem clearly is arc welding. This process usually requires several separate items to be fitted into jigs, prior to the welding being performed. Whilst it may be technically possible to load the jigs automatically, the incurred cost often outweighs any manpower savings that could accrue. Further, automated loading of anything but the simplest item can often take far more time than that required by a human loader.

The man-robot solution, in which the operator is primarily occupied with loading and/or unloading the workstation, requires two peculiar issues to be resolved if it is to function effectively:

1. The jigging that holds the components being processed must be designed and manufactured so that the presented components are held firmly in a unique position relative to the robot. At the same time, the jigs must also be capable of easy manual loading and unloading without recourse to elaborate tooling or contortions. This precise jigging is needed because of the 'low intelligence' of the present day industrial robot systems, whose programming expects and requires a deterministic environment in which all features of the workstation are described. Whilst a person can perform selective assembly, or put in an extra weld to fill up an extra wide gap between components, which has been caused by distorted components or poor jigging, the robot can only perform what it has been taught.

2. The system must be interlocked so that the shuttle or rotary table, used to present the work to the robot, does not move until initiated by the operative and authorized by the robot control system. This means that the system is paced by the loader/unloader, with the added benefit that the risk of injury to the operative from unexpected movements of the system is minimized.

Programming

Programming is one of the most important tasks involving the direct interaction of humans with robots. Even if the program is developed offline, it still has to be proved, for safety reasons, with the robots on the assembly line. The reason is that, unless the system is known totally as a quantitative 'equation', there will be areas which are known only in vague terms of knowledge. For example, the exact relationship of the different axes of the robot to each other, for different extensions or loadings, might not be known. Hence, the offline program has to assume that all the relationships are square and parallel, and that the deflections are 'negligible'. It must be appreciated that such assumptions could have disastrous results in a complex system.

It makes economic sense to use humans for programming, maintenance, and other tasks that require judgement. The advantages are low cost, a human's adaptability and innate intelligence. Disadvantages are:

1. The 'unreliability' of humans as productive elements.
2. Any psychological problems experienced by any member of the workforce in close and frequent contact with the robots.

PSYCHOLOGICAL PROBLEMS

Major psychological problems are usually generated by the need for a programmer to think in terms of how a task can best be performed by the robot. If the program is entered without any thought to the robot's capabilities and control system, then often the program replay can seem awkward, or the robot appears to have a 'mind of its own'. Consequently, there is a lot of program debugging and fine tuning to give the robot the correct smoothness of operation.

Alternatively, if a human is working with a robot, then apart from monitoring the performance of the robot, he often has to perform a separate task. If automatic monitoring equipment is installed, the supervisor might wonder which of the two workers is really being monitored – the robot doing the task, or the supervisor on the basis of how well the task is being done.

Humans can also become alienated from the role of programmer or coworker, because the robot does not communicate socially with its human counterpart. If human skill is not practised, it degenerates. Hence, if most of an operator's time is spent overseeing a robot, it follows that when a manual task is called for it is unlikely to be performed well. Thus, a human's neuromuscular skill is not only usurped by the robot, it also deteriorates, which may lead to resentment. This skill deterioration does not of course occur with the robot. Once a pattern is programmed, the information is stored on either a battery backed, or nonvolatile, memory. Therefore, even if the main power is switched off, the information and skill level of the robot remain until deliberately altered or erased. Further, because robots are faster, more accurate and more reliable, it is a human failing to attribute qualities to them that they do not have (eg wisdom, judgement and worthiness). Hence, the robot's personality can be both seductive and alienating.

Man-robot systems

Man-robot systems are not peculiar to, or limited to, arc welding applications. They occur whenever it makes technical or economic sense to use a human to perform certain tasks required by that workstation. In many cases where, for instance, the workstation is processing a number of similar components or products, it is the human that 'identifies' the items and, by overt or covert selection, 'informs' the robot what program to use for that particular item; for example, the assembly of castors in which different handed or sized items require the same technical task performed, albeit at different geometric or dimensional positions. The operator loads the appropriate handed jig which, on arrival at the robot's work zone, activates coded instructions through its own built in electronic sensors. This is not to say that a robot system could not recognize components, merely that the use of nonhuman sensors for initial identification could entail a technically complex interface and might not make economic sense for that application. Further, because of preassembly processes, the components are presented to the system in a disordered state; they can of course be easily accommodated by the human. However, it does not make industrial or economic sense to use robots to cope with a disordered environment (eg bin picking). However, this does not mean that disordered presentation of items is a valid reason for retaining operators, or that everything should be presented in an ordered manner. It merely means that sometimes there is no rational or economic procedure for ensuring ordered presentation.

Another example is where the product being assembled contains a number of soft items (eg foams or electrical wires). The handling of many of these soft items does not present any insurmountable technical challenges, but the cost of developing the solutions cannot always be justified. Alternatively, an

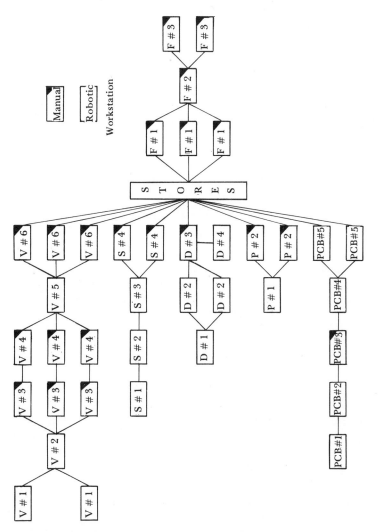

Figure 11.1 *A workstation allegiance of a 38-workstation flexible assembly system (FAS). The FAS has workstations in a parallel and serial arrangement so that line balance is achieved. Components are processed in five work areas (PCB, P, D, S, V) and are then placed into the stores prior to making into products against orders in the sixth work area coded 'F'.* (from *Flexible Assembly Systems*, Plenum Press, 1984)

assembly task might require that a particular component be inspected prior to assembly for blemishes or other surface defects. Again, the use of people to perform this task makes

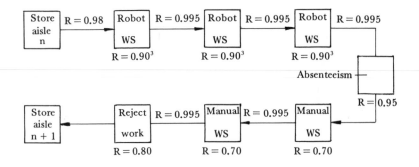

$$R_{\text{work area}} = 0.98 \times 0.90^9 \times 0.95 \times 0.70^2 \times 0.995^5 \times 0.80$$

$$R_{\text{wa}} = 0.138$$

Figure 11.2 *Reliability of mixed man-machine lines can be computed as shown. The validity of the computed reliability depends upon the data used, hence care must be used when selecting reliability values to each element within the system.*
(from *Flexible Assembly Systems*, Plenum Press, 1984)

far more economic sense than developing an automated solution.

Figure 11.1 shows a workstation layout for a flexible assembly system, developed for the processing of money changing machines. For assembly reasons similar to those mentioned above, it is necessary here to use a mix of manual and robotic workstations.

If humans are used as direct productive elements within an assembly process, then inevitably they affect the productivity of that line (eg through absenteeism, variance of output and variance of quality). As with any system, provided that quantitative values can be given for these variables, the productivity of a man-machine line can be computed. The example given in Figure 11.2 assumes that the robot workstations had reliability values of 90%, whilst the values for manual workstation reliability, attendance and acceptable quality work were 70%, 95% and 80% respectively. The reliability of the material movement between the stores and the various workstations does not necessarily mean that the equipment has broken down, only that it cannot service the next 'module'.

CHAPTER 12

Safety

A robotic assembly system is an integration of robots, machines, computerized information channels and humans, no element of which can be considered perfect or immune from eventual failure and malfunction. The proximity of humans to the robots allows the risk of mutual damage, resulting in the formulation of safety guidelines that indicate how the conditions of conflict can be minimized. The high productivity levels associated with robotic systems can only be realized if all the system elements are functioning safely and reliably.

However, until definitive regulations are issued by law, safety comes down to common sense and the application of the 'what if . . . ?' syndrome. Attempting to determine the safety hazards of a robotic assembly system is best done on a piecemeal basis, whereby each element is analysed for risk. The relationships between each element are known either on a quantitative or, at worst, a qualitative basis. Therefore the risk factors can be transferred from one element through to the others. If a fault tree analysis is used, then the relationships for various associated activities can be shown logically. Each element and its probability value can be plotted on an event tree to determine the risk of a number of events occurring, so causing a safety hazard.

Humans at risk from injury by robots
There are four groups of humans at risk from direct personal injury by a robot:

1. Programmers.
2. Maintenance engineers.
3. The casual observer.
4. Others outside the assumed danger zone.

PROGRAMMERS

A robot programmer, using any of the three online programming techniques outlined in Chapter 3, is in direct physical contact with the robot. This closeness with the robot's work envelope, with its inherent dangers of injury or death, distinguishes robotics from any other form of automation.

Although it is acknowledged that robot motions during programming are usually only a fraction of their functional operating speeds, malfunctions do happen *and* it is possible for an inexperienced programmer to get into trouble through interest in the process and ignorance of what is happening to the robot's appendages and associated equipment. For instance, a programmer could be in a position such that the robot hits him as it moves from one point to another. This causes the programmer to make a random pendant instruction (in panic) that activates the robot's process program.

Fortunately, 'assembly robots' are relatively small and the hazards of programming robots for assembly processes are generally limited to being trapped between the robot and another object within the workstation, or being bruised by the robot's arm. This is not to say that the programmer should not be careful, since a sharp edge on a gripper, or a long thin component being manipulated, could easily rip a hand or face, or even blind the programmer.

MAINTENANCE ENGINEERS

Maintenance engineers are at risk from much the same dangers as programmers, with an added risk of electrocution. Also, because maintenance procedures often require that safety interlocks are disconnected, the inherent risk of injury is greater.

THE CASUAL OBSERVER

To the casual observer, robots are often seen standing still, apparently doing nothing, for long periods of time. The programmer would of course know whether or not these pauses are intentional, with the robot performing a programmed delay or waiting for an input from another element of the workplace so that it can continue its task. However, if as is usually the case, the assembly robot is not rigidly guarded, then a casual observer, if so inclined through curiosity, may move towards a 'stationary robot' and be injured when the robot continues its operation.

OTHERS OUTSIDE THE ASSUMED DANGER ZONE

Even though a robot has a known maximum work envelope, the risk of injury is not limited to encounters within this envelope. If grippers, or components manipulated by the assembly robot, are not properly designed, then it is possible for the components to be flung out of the grippers, due to inertial forces, and strike personnel well outside the assumed danger zone of the robot.

Safety procedures and devices

In the practical sense, safety procedures and devices allow the authorized entry of humans into a robot's working envelope, with the minimum risk of injury. Unauthorized entry into an active robot cell should always be discouraged. When activated, the robot's cell 'guardians' close it down either immediately or after completion of a 'present action', the option being dependent upon the sensor activated. Hardware devices (eg fences, pressure mats, etc) and sensors monitor and guard all anticipated reasonable access to a robot's work envelope.

THE EMERGENCY STOP

An emergency stop (ie immediate total shutdown) normally occurs after a collision, which necessitates the robot being put in null balance so it can be moved away from the collision point and/or the trapped 'item' released. The direct startup afterwards is dangerous because on certain robots the controller may not know where the robot is in terms of both the program it was following and its present position. Consequently, start-up causes the robot to make random actions in an attempt to continue the program it was performing, resulting in possible damage to the robot, workstation or product.

CONTROLLED SHUTDOWN

An alternative to the emergency stop is the 'controlled shutdown'. Here, the robot finishes the action it was performing, prior to the shutdown command. Thus, the program may be restarted without concern over random moves or damage. This option is identical to single stepping through a program, or activating a pause in the program.

Obviously, on procedural and access grounds, this type of shutdown has to be examined for potential hazards, especially if close proximity access by the operative to the robot is permitted.

PHYSICAL SAFEGUARDS

Physical safeguards are many and varied. For instance:

1. Simple contact switches, through to magnetic locks.
2. Restrained keys.
3. Pressure mats to infrared light beams and vision systems.
4. The use of a flashing red light within a work zone will indicate that an apparently stationary robot is activated, but awaiting an input, or performing a time delay.

Collision sensors

Many robots are fitted with 'collision sensors' on the sides of their arms. In the event of a collision, contact is closed and an emergency stop results. Proximity sensors can also be used on robot arms or grippers in multirobot situations and used for procedural checks to prevent collision.

Caged robots

Robots can be caged to allow simultaneously visual checks by the operator, authorized access through an interlocked gate, and to prevent flying debris from exiting from the robot work space.

In complex automated systems, where there are 'lines of robots', each robot can be separately caged so access to any one robot for programming or maintenance reasons can be safely achieved without the need to shut down the entire line.

Procedural checks

In the wider sense, safety refers to more than just preventing harm to persons in the vicinity of the robot. It covers the prevention of damage to the robot itself, as well the knowledge that the danger zone (for employees) can extend beyond the confines of the workstation. The robot's own I/O ports allow it to check its input signals and, after making each action, to send output signals to inform other elements within the 'cell' that the action is completed.

The use of procedural checks and the sequencing of the robot's actions to those of other machines, and/or persons in its work envelope, minimize the risk of collision. If, for instance, a press and robot are not synchronized by interlocks, the robot may attempt to enter the dies of the press before the dies are open, or the press may close while the gripper is between the dies. In both cases, the gripper or dies could shatter and cause flying shards of metal to enter the general work space. Additionally, the 'accident' would bring the assembly line to an entire stop, incurring two economic penalties: one for repairs and one for lost revenue until the assembly station is repaired.

CHAPTER 13

Evaluation of a robot system

Technical and economic reasons are the primary reasons for using robots to perform assembly tasks. Social issues are a secondary incentive, and all these three items have a financial dimension.

The economics of an assembly process are based upon it producing items at a predictable rate and cost per unit. The process must also meet a number of technical criteria if it is to yield products of a consistent quality and high functional reliability. The financial success of a product is measured by the revenue it generates, which in turn is governed by the marketplace acceptance of that product. Productivity and quality assurance of the assembly process are related to the suitability and reliability of the technology incorporated, which shows itself in the prime cost of the items assembled.

Objectively, manual and robotic assembly systems can be compared by five parameters: capital cost, output, quality, reliability and cost per unit:

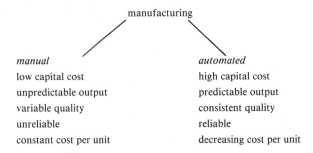

manual	automated
low capital cost	high capital cost
unpredictable output	predictable output
variable quality	consistent quality
unreliable	reliable
constant cost per unit	decreasing cost per unit

There is obviously no benefit in exploring the use of robots for assembly if there is no financial benefit to be gained. Profit is the most important financial measure, and it can be expressed thus:

Contributions to profit = selling cost − prime cost

Selling cost is set by market forces, so contributions to profit depend upon the prime cost, which in turn has several components:

Prime cost = material & components + direct labour + processing + overheads

scrap	wages	machining	supervision
inspection	social cost	assembly	rates
sorting		waiting	management
storage		capital cost	commercial
purchasing		maintenance	other indirects
delivery			

Another way of looking at prime cost is through the following equation:

Prime cost = cost of material + value added

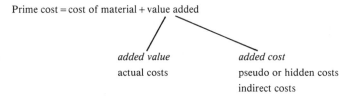

added value	added cost
actual costs	pseudo or hidden costs
	indirect costs

Therefore, to maximize contributions to profit, the following should be reduced:

1. Labour content.
2. The hidden added costs.
3. The scrap.
4. The variation in quality.
5. The storage costs; make goods against orders, not for stock.
6. The per unit cost.
7. The consumable materials used, which should also be controlled.

It is not enough just to have a check list of the factors that need to be taken into account when making a proposal. What is needed is quantitative data about all these factors, so the proposal can be substantiated objectively. Figure 13.1 shows how profitability is directly related to, but not guaranteed by, quantitative information. The relationship is genuine and valid, since 'real' knowledge allows the assembly system to be understood and, if necessary, computer modelled. Thence,

153

control can be implemented which, if correct, will lead to increased efficiency with its corollary of better productivity. This, in turn, permits profits, assuming that the factors pertaining to market, cost and selling price have been similarly analysed.

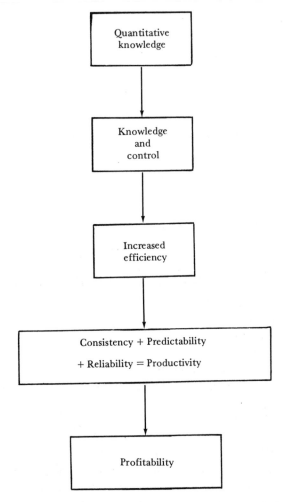

Figure 13.1 *Profitability is the ultimate reason for using robots to perform assembly tasks. The level of profitability is a function of the productivity of the system, which is itself directly related to the level of quantitative information, as well as to the application of monitoring systems so that the assembly systems exhibit acceptable levels of consistency, predictability and reliability for the product being processed.* (from *Flexible Assembly Systems*, Plenum Press, 1984)

Methods of financial appraisal

There are several methods of evaluating proposals for capital expenditure. It is important to realize that the choice of the appraisal method can restrict or even distort expenditure programmes. The methods discussed here are:

1. Payback period.
2. Return on investment.
3. Discounted cash-flow.
4. Internal rate of return.
5. Life cycle costing.

PAYBACK PERIOD

This is the most simple and commonly used method of analysing a robot proposal. It involves computing the time taken to recover the initial capital outlay, and is normally expressed in years.

The computation is usually easy and, since the time scales are normally short (ie usually two years), little account is taken of the implications of interest rates, or inflation. In uncertain economic conditions, this method certainly provides a measure which reduces the risk when short payback periods can be shown. However, if used as the sole appraisal method, it can lead to the wrong decision being taken:

1. It ignores income after the payback point.
2. It cannot indicate the relative profitability of a project.
3. It tends to concentrate on limited variables (eg labour savings), whilst ignoring less significant, intangible benefits.
4. It is biased against investments which do not yield their highest returns in their early years.

RETURN ON INVESTMENT

Return on investment (ROI) is another comparatively simple method of performing project appraisal. It uses the ratio of net profit to capital employed. Profit is normally averaged out over the anticipated life of the installation and no account is taken of the time value of money. This method has serious limitations, since it discriminates against projects of less than ten years. It is also unsuitable for the task of optimizing investment decisions.

155

DISCOUNTED CASH-FLOW
Discounted cash-flow (DCF) techniques are designed to take into account the importance of the time value of money (ie a unit of currency today does not have the same value as a unit of currency presented at some future date). There are two major applications of the DCF principle:

1. The net present value (NPV).
2. The internal rate of return (IRR).

NET PRESENT VALUE
The net present value is considered to be generally the more realistic, and calculations are used to assess whether a project will provide a negative or positive return on capital (expressed in units of currency). The flows of money are discounted during the period, using a rate specified by the company.

INTERNAL RATE OF RETURN
Internal rate of return is closely related to ROI, but is an attempt to determine the real discounted rate of return, which can then be assessed to see if it is acceptable within corporate objectives.

LIFE CYCLE COSTING
Even when the time value of money is taken into account, through the use of discounting techniques, the data is limited to certain aspects of capital projects. To counter this restricted approach, the use of life cycle costing (LCC) has and is being encouraged. This procedure takes into account *all* cost factors from the date of the decision to acquire the asset, right through to its disposal.

The UK Department of Industry's (1977) definition of life cycle costing was:

> 'LCC includes the costs associated with acquiring, caring for and disposing of assets; feasibility studies; research; design; development; production; maintenance; replacement and disposal; as well as support, training and operating costs generated by the acquisition; and the use, maintenance and replacement of permanent physical assets.'

The benefits of such an approach are particularly relevant to investments in robots, because the robot's flexibility ensures that it retains its usefulness as a productive element, long after the demise of the product for which it was purchased. Apart from indicating the anticipated costs of using robots over a protracted time scale, life cycle cost investigations will highlight the resource implications of such a decision, and be more applicable to revealing the financial advantages offered by more flexible systems than are the more traditional appraisal methods.

Although offering a much more thorough approach to project appraisal, LCC makes great demands upon the resources of an organization. However, it does present management with a greater comparative understanding of the economics of alternative proposals and can certainly expose the weaknesses of adopting the 'lowest bid' without taking into account all the longer term costs. In other words, it shows the trade-offs of low capital costs with high lifetime operating costs (manual processes) *versus* a high initial capital cost and low operating costs (robotic processes).

Strategic and tactical justification

The guidelines for robot justification are that the proposal should deal with both the short-term tactical reasons and the long-term strategic advantages of adopting robots for an assembly task.

Tactical economic justification is concerned primarily with the discrete units of capital investment required for the next six months to three years, with the justification being driven by short payback periods. The main emphasis is on cost analysis, cost avoidance, reaction to product changes, improvements in quality, the levels of scrap and rework, and short-term marketing, or customer demand cycles.

On the other hand, strategic justification takes into consideration the following:

1. The long-term benefits of introducing a new technology.
2. The market forces that will cause changes in product mix.
3. Workforce characteristics.
4. The demographic patterns of the entire manufacturing industry.

Strategic justification investigates how a number of robot systems could be tied into a group of higher level control systems to provide a communication-based automated production line. Alternatively, it can investigate the implications of interfacing the robots with other computer integrated manufacturing systems, thereby deriving benefits from the systems approach.

Strategic considerations may include:

1. The potential for additional business opportunities.
2. The need for the adoption of robots so that a competitive position may be obtained or retained.
3. The overall equipment and plant modernization strategies.
4. Provision for high technology advances to ensure the retention of a strategic position by a company or industry.

Justification factors for robots are listed in Table 13.1.

Strategic	Tactical
Technological advancement	Productivity increase
Competitive position	Cost avoidance
Investment capital for growth	Quality/scrap/rework
Modernization	Reaction to product changes

Table 13.1 *Justification factors for robot use*

Productivity ratios

Productivity is considered to be the most important, if not the key, measurement for the efficiency of a system. Yet there is not a single, unique definition for productivity. Productivity ratios used vary from country to country, industry to industry, and factory to factory. The measures used are all subjective, their validity varying according to the industry sector and production class. Sutton (1980) claims that it is necessary to use substitute or surrogate production ratios that relate to key parameters within the sector/class of interest. This is similar to the analysis ratios used by accountants to compare objectively a company's performance against others in the same industrial sector. They can also be used to compare different divisions within the same company, so that corrective actions can be applied, where necessary, and the results highlighted.

The values and indices listed in Tables 13.2 and 13.3 are neither absolute nor exclusive, since different assessment criteria may require new or unique ratios to be constructed. What is certain is that, if the ratios are to be of benefit, the information used for assessment must be up to date, relevant and valid.

Robot versus manual cost per hour

A common focus for management when preparing proposals for robot implementation is the reduction of direct labour in the assembly process. In many proposals, this is one of the few, if not the only, measurable and tangible cost savings. It is beneficial to compare the *actual* costs per 'productive hour' of both humans and robots.

A direct labour worker is normally paid on an hourly basis which, in the USA, averages at present about $8 per hour, plus between $4 and $7 per hour in associated benefits and costs to the employer. The typical cost for an eight-hour shift is therefore between $96 and $120. In many companies, however, the actual time a manual worker is performing value added work during a shift can be as low as 6½ hours. If we normalize the actual hours spent in productive work and the daily rate, an hourly rate of between $14.76 and $18.46 is obtained.

Assuming a $40 000 robot has a design life of 20 000 hours (which is ten years at a single shift), then its hourly cost over that time period would be $2 per hour. If it is anticipated that a robot will become obsolete within one year, taking into consideration a double shift operation, a cost per hour of $10 is derived. This raw comparison does not include tax advantages, the cost of borrowing money, the depreciation of the robot, or the lowered value of any money gained at some future time.

Figure 13.2 shows a comparison and deviation of hourly cost between humans or robots, which also take into account many factors such as maintenance, power, installation, overhaul, depreciation. It also assumes that the robot is not discarded once it becomes technically obsolete for its original task.

In brief, then, it is a mistake to assume that the cost of manual direct labour is about $6 per hour, since taking all the different direct and indirect costs into account, the actual value can be as high as $50 or $100 per hour. Hence, the one-to-one comparison of robots to direct labour is as obsolete as comparing logarithm tables with modern day calculators.

Resource graphs

When making a proposal, the first fact that has to be determined and specified is, what is the product to be processed,

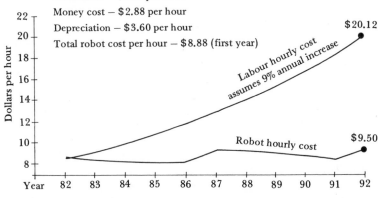

Robot hourly cost comparison

Robot cost — $50 000
Useful life (for depreciation) — 8 years
Anticipated usage — 2 shifts
Cost of money — 20%
Installation cost — $10 000
Maintenance cost — $1.05 per hour
Overhaul (twice in 10 years) — $1.00 per hour
Installation — $1.20 per hour
Money cost — $2.88 per hour
Depreciation — $3.60 per hour
Total robot cost per hour — $8.88 (first year)

Figure 13.2 *Starting with the robotic and manual hourly cost being approximately the same in 1982, this figure shows how the cost curves diverge rapidly over one decade. The curves are based on the argument that a robot has a long useful lifespan and that, provided its purchase and installation costs are amortized over that lifetime, the cost per hour will decrease over time. On the other hand, if the designated tasks are performed manually, the incurred production costs will increase yearly. Although the curves can be challenged at a detailed level (because the data used is fixed in time), the general prediction cannot be faulted.*
(from Robotized multicells best for mid-volume output,
Robotics Today, April 1983)

and why is a robotic system needed to assemble it? The proposal must be product-based, since if there were no product there would be no market to be satisfied. Hence, if there were no (potential) revenue, why would a robot system, or any other assembly system, be required? This is not to say that a product must exist as a proven entity before a robot system can be considered; rather, that the product to be processed must be definable in general technical and marketing terms.

Any product requires a given number of hours to assemble. It also has an annual demand (that is known or projected). If these two factors are combined, the time required to assemble a number of products would result. A year is a finite amount of time and, in terms of the capacity of the assembly system, is a function of the number of personnel and/or robots available to satisfy the demand for that product. If there are a number of different products, then it is possible that there could be some conflict of resource demand if the system tries to satisfy the output demand for all of them.

$$\text{Production index} = \frac{\text{Recorded production hours}}{\text{Total scheduled hours}}$$

$$\text{Availability index} = \frac{\text{Total scheduled hours} - \text{recorded maintenance hours}}{\text{Total scheduled hours}}$$

$$\text{Utilization index} = \frac{\text{Recorded production hours} + \text{recorded maintenance hours}}{\text{Total hours in time period considered}}$$

$$\text{Cycle cutting ratio} = \frac{\text{Machine cutting time}}{\text{Total cycle time, including setup, loading, and tool change}}$$

$$\text{Expansion ratio} = \frac{\text{Scheduled production hours}}{\text{Maximum possible production hours available}}$$

$$\text{Maintenance ratio} = \frac{\text{Maintenance cost of machines}}{\text{Machine hours available}}$$

$$\text{Output per hour} = \frac{\text{Number of parts manufactured}}{\text{Machine hours operated}}$$

$$\text{Rejection ratio} = \frac{\text{Number of parts rejected}}{\text{Total number of parts manufactured}}$$

$$\text{Removal index} = \frac{\text{Amount of material removed (weight)}}{\text{Total cutting time}}$$

$$\text{Tooling ratio} = \frac{\text{Cost of tools and fixtures}}{\text{Total production cost}}$$

Table 13.2 *Substitute productivity ratios for automated production equipment*

If a graph is drawn of hours needed per assembly *versus* annual demand, then a limit can be drawn that equates with the maximum number of products of a given assembly time (per unit) that can be processed by one person. The boundary can be extended to compensate for 'n' persons and/or two or three shift workings. Projection of the assembly time and the annual quantity of a particular product will intersect at a unique point on the graph. If this point is within the boundary, processing

Employment ratio =	$\dfrac{\text{Value added}}{\text{Number of persons employed}}$
Energy ratio =	$\dfrac{\text{Number of units manufactured}}{\text{Energy used (kWh)}}$
Work ratio =	$\dfrac{\text{Employee hours worked}}{\text{Employee hours paid}}$
Material ratio =	$\dfrac{\text{Material in final product}}{\text{Total material supplied}}$
Quality ratio =	$\dfrac{\text{Total cost of quality assurance}}{\text{Total product cost}}$
Cost per part =	$\dfrac{\text{Total operating cost}}{\text{Number of parts produced}}$
Capital intensity =	$\dfrac{\text{Capital equipment employed}}{\text{Number of human employees involved in process}}$
On-time delivery ratio =	$\dfrac{\text{Number of on-time deliveries}}{\text{Total number of deliveries}}$
Inventory turnover =	$\dfrac{\text{Cost of goods sold}}{\text{Average inventory for the period assessed}}$
Recycled material ratio =	$\dfrac{\text{Recycled material used (weight)}}{\text{Total material used (weight)}}$

Table 13.3 *Figures of merit of indices used in lieu of productivity ratios for evaluating industry and/or company productivity changes*

by a manual team is possible. Conversely, the point will indicate the number of persons and/or shifts needed to satisfy (just) that product's demand. This will permit management to calculate the potential costs of satisfying that order, and so make decisions regarding the *value* of alternative orders using the same resources.

Figure 13.3 shows a resource graph where a given combination of assembly time and annual demand falls either inside or outside the region enclosed by the ordinates and the boundary. The graph can also be used to examine the potential manpower saving that could be realized through automating some of the processes. A very crude method of assessing the return for a particular investment is to consider manpower saving alone:

$$\text{RoI} = \frac{n(\text{annual cost of operator})}{\text{cost of automating alternative}} \times 100$$

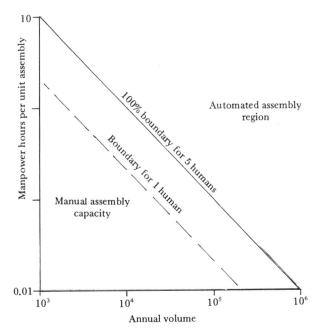

Figure 13.3 *The resource graph shows the capacity of one to five humans for a given volume of product. The graph boundary can be varied outward for additional humans, or inward to take account of the inefficiencies of manual productivity. By plotting the annual volume and the work content of a particular product, it can be determined whether the available human resource can accommodate the workload, or whether it falls into the region identified as that of automated assembly. This external region merely indicates a larger capacity than that within the boundary. It can therefore be encompassed by increasing the personnel level in the assembly tasks, or by using an assembly system that has a larger output per unit time.*
(from *Flexible Assembly Systems*, Plenum Press, 1984)

This approximation does not take into account any of the benefits achieved from an alternative solution in terms of more consistent quality and predictable output; nor does it take into account depreciation, inflation or the discounted value of money.

If more than one product is to be processed, to determine the capacity needed to satisfy the demand for all the products within a given time frame, it is necessary to weigh the various processing times against the total volume being processed. For n products, the composite processing time per assembly is:

$$T_n = \sum_{i=1}^{n} \frac{V_i}{V_n} * t_i$$

where T_n is the total hours, V_i is the volume for the ith product, V_n is the total volume of all the products, and t_i is the processing time for the ith product.

It is also important to identify the correct process rate for the product(s) per unit time. Apart from the pure assembly tasks, there is the time required to set up the workplace so that the products can be processed, as well as the time used for inspection and workplace servicing. If the value allocated as the boundary is the maximum possible number of hours per 'n' man shift year, the graph will be in error because it ignores many of the factors that can limit an effective working year.

PLOTTING MANUAL ASSEMBLY

If manual assembly is considered alone, the 'agreed' processing rate incorporates a factor that takes into account operator fatigue and variation of output. This factor usually increases the time permitted to do a given task by about 20%. In addition, if the operator relies upon automated or semi-automated systems to transfer materials to and from the workplace, there is a high probability that delays could occur. In consequence, an operator may not be able to assemble the products at the assumed rate. Likewise, illness or natural breaks, union meetings, and other activities, reduce the time available for the assembly process. Hence, the limiting boundary should be adjusted to reflect the real conditions of the system.

PLOTTING AN AUTOMATED PROCESS

If, in Figure 13.3, a product is plotted outside the manual boundary, one option would be to use an automated process to assemble that product. Again, a resource chart can be drawn (Figure 13.4) that indicates the capacity of the automated alternative. The factors that limit the boundary of the manual process can be used to show that the capacity of an automated system is between 1.25 and 2.50 times the *potential* capacity of a manual system. The capacity of the automated system is calculated from the number of days available, the number of shifts per day used, and the expected uptime for the system. However, this maximum must be reduced by the time required for regular maintenance and service, as well as the time needed for any extra programming (Owen 1984).

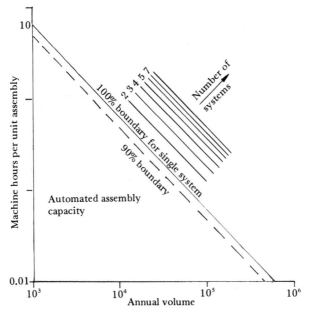

Figure 13.4 *The resource graph for an automated assembly system. Here the boundary can be increased by varying the number of systems, or decreased by taking into account the downtime of those systems, and/or the number of shifts operated.*
(from *Flexible Assembly Systems*, Plenum Press, 1984)

Cost groups

Before a full comparison can be made between one assembly system and an alternative (see also Chapter 14), it is essential that the full and total costs for each system are listed. There are three cost groups that have to be considered:

1. Fixed costs.
2. Variable costs.
3. Hidden costs.

FIXED COSTS

These are the overheads that are apportioned by management to the system, the depreciation of the equipment, and all other costs that do not vary as a function of the output of the system.

VARIABLE COSTS

These are the directly incurred costs that are experienced in the assembly of each and every product that is processed by the system. Typical of these costs are material, labour (if applicable), energy and other expendable costs that vary directly with the output of the assembly system.

HIDDEN COSTS

These costs tend only to be assessed (if at all) in qualitative rather than quantitative terms. Ignorance or omission of these costs will be detrimental to the viability of a proposal for the use of robots for assembly. Examples of these hidden costs are:

1. Scrapped materials and energy.
2. Wasted labour incurred through scrapped items.
3. Lost sales revenue due to scrapped products.
4. Costs incurred in refurbishing/replacing products that have failed within the guarantee period.
5. Costs incurred in reworking or salvaging inferior quality items.
6. Costs of inspection facilities because of low competence levels of manual operatives.
7. Lost sales revenue because of inefficient production methods resulting in low throughput.
8. Lost opportunity costs because of monies tied up in inventory as a cover for fluctuating production levels.

TANGIBLE AND INTANGIBLE COST FACTORS

Quite often, when a proposal is being made up, it is implicitly assumed that everything in the working environment is known deterministically and has been accounted for. An indication of some of the tangible and intangible cost factors involved in an assembly process are given in Table 13.4.

Tangible factors (mainly techno-economic)	Intangible factors (mainly socio-organizational)
Cost of machines	Cost of quality control
Cost of jigs and fixtures	Cost of supervision and manpower control
Cost of power supply	Cost of absenteeism
Cost of labour	Cost relative to job satisfaction
Cost of maintenance	Cost due to grievances or strikes
Cost of setup	Cost of organizational flexibility
	Cost of occupational disease

Table 13.4 *Tangible and intangible cost factors*

Other factors that have a dramatic effect upon the cost of an assembly system are:

1. Useful life of the system.
2. Number of pieces required per unit time.
3. Time to assemble a specific number of items manually.
4. Time to assemble a specific number of items robotically.
5. Number of executed operations per assembly.
6. Life cycle of product.
7. Number of product types for which system was designed.
8. Functional, geometric and power parameters of the system.

The values and influences of the vast majority of these factors cannot be known deterministically until the assembly system has been installed and operated. To assume otherwise would result in an erroneous conclusion being drawn from a proposal.

The proposal

In order that the proposal is as complete as possible, it is necessary to determine which, if any, factors are more influential than others, and then to allocate values to them. When a factor is not known in quantitative terms (deterministically), it is often known in qualitative terms, whereby it can be said that a certain situation is liable to happen under a given set of circumstances. An allocation of numerical values to these qualitative terms and conditions is known as stochastic or probabilistic analysis.

The decision to use robots to perform a given assembly task should not be made on the sole basis of technical viability. The reason is that, like any other project, the adoption is going to require capital investment by the company. Therefore it should not be considered in isolation, or as a special case, but be treated as any other project whereby a well constructed and documented proposal listing the technical, economic, financial and social reasons for the adoption is made to management.

It is advisable that the information is given in a format that is objective, and that does not cloud the issues with opinions, preferences or subjective issues. Failure to put forward a comprehensive proposal could result in the project being rejected out of hand, irrespective of the amount of work that has been done to prove its technical viability.

It is important to remember that a proposal will invariably be judged against a financial rather than an engineering background. Hence, it is imperative to understand the different views of accountants and engineers *vis à vis* what constitutes a viable project. The engineer will be looking at the technical merits of the proposal, whereas the accountant will be assessing the manpower saving.

However, in the end, justification for using a robot for assembly tasks comes down to carefully weighed arguments, in both the engineer's and accountant's languages, which will make or break the proposal. This problem of language is particularly relevant to proposals involving the use of robots, since the specifications (and quotations) are written by those familiar with the robotics industry and are fluent in the 'lingua robotica' (Owen 1984). Unfortunately, many decision-makers are not aware of the existence of this 'language' and consequently differences do occur between expected and real performance.

CHAPTER 14

Economics of alternative systems

When comparing alternative assembly systems, it is important to know what is being compared and against what criteria. For example, when comparing operating costs, it is necessary to ensure that it is for the same quantity, or over the same time period. The determination of an alternative assembly system can be made on the basis of the following:

1. Automating an existing system without necessarily wanting to increase the production quantity.
2. The comparison of two automated systems to produce a given quantity of products.
3. The cost per unit across a range of quantities of a given product.

Assessing costs

Comparing the costs of processing a product by one of the three alternative methods (ie manual, flexible automation or hard automation) is the subject of many research and conference papers. One research project used the product of the system's assembly time and the cost of the system to give a value in dollar-seconds. In this way, the dollar-second value for each alternative can be determined. Then, irrespective of the particular values of time or cost, if the price-time product is the same, it does not matter which system is used. Obviously, this is on the assumption that one or other of the two factors does not rule a particular system out because of other criteria (Nevins and Whitney 1979). For instance, if one system cost $100 000 and it took two seconds to assemble an item, and the second took four seconds, but only cost $50 000, both would have a price-time product of 200 000 $ sec. The choice would then have to be made on other criteria, eg reliability or future demand predictions. Using typical values of system components, labour costs and material, etc, the same research has determined that for a robot system to be economically viable against either manual or hard automation alternatives its price-time product must not exceed 293 000 $ sec.

Equilateral triangle

An equilateral triangle, with logarithmic scales on all three sides, may be used to determine the economic advantage between alternative assembly methods. The basic equation of the triangular relationships is (Csakvary 1981):

$$\frac{C}{n} = Mt$$

where C is the cost of assembly, M is the cost of operating an assembly station per unit time, n is the number of parts assembled to the product, and t is the average assembly time per part assembled in the product.

While this might seem a very simple relationship, in reality it does include 22 variables. The equations are also modified into three forms, one for each assembly alternate:

1. Manual assembly:

$$C = kt_M(1 + X)\left[\frac{n}{k}W\right] + \frac{nW}{kSQ}(2C_M + YC_P)$$

2. Dedicated assembly:

$$C = (t_D + nXT)\left[2W + \frac{nW}{3k}\right] + \frac{(Y + n)W}{SQ}(C_T + C_P + C_W + C_{DC})$$

3. Programmable assembly:

$$C = \left(\frac{k}{2}t_p + kXT\right)\left(2W + \frac{nW}{3K}\right) + \frac{nW}{kSQ}(C_T + C_P + C_R + C_{PF} + C_{PC})$$
$$+ \left(\frac{(Y + n)W}{SQ}\right)C_G$$

where C is the assembly cost per product assembled; C_{DC} is the cost of a dedicated station computer; C_F is the cost of an automated part feeder; C_G is the cost of a gripper; C_M is the cost of a transfer device per manual assembly station; C_P is the cost of a work carrier; C_{PC} is the cost of a programmable station control computer; C_{PF} is the cost of a programmable part feeder; C_R is the cost of a robotic or programmable work head; C_T is the cost of a transfer device per automatic assembly station; C_W is the cost of a dedicated work head; k is the number of parts assembled at one station; n is the total number of parts assembled per product; Q is the equipment equivalent cost of an assembly operator; S is the number of shifts per day; T is the machine downtime due to defective parts; t_D is the average assembly time per part (dedicated); t_M is the average assembly time per part (manual); t_p is the average assembly time per part (programmable); W is the manual assembly labour rate (including overheads); W_m is the total cost of manual assembly (including overheads) which equals $(n/k)W$; X is the ratio of defective parts to acceptable parts, and Y is the number of product styles, as a function of assembly.

The annual volume is plotted on the base line of the triangle and a perpendicular line drawn. The number of product styles is then plotted on the righthand side of the triangle and another perpendicular line drawn to intersect with the first. From this intersection, a third line is drawn perpendicular to the lefthand side of the triangle. This last line passes through the three parallel lines that represent the relative cost of assembly per product produced. Where one automated assembly method dominates strongly over others, only that scale will be 'marked'. The scale representing manual assembly is there for two purposes: to provide a base for converting from relative to specific cost, and to provide a reference line so that, once the selection has been made between the two alternatives, the economics of the selected method to that of the (existing) manual method can be performed.

Figure 14.1 shows the selection triangle marked up for two test cases. The first, (1), (2) and (3), show that the flexible automated option is the best choice, costing approximately 0.6 times less than if the task were performed manually. The second case indicates that dedicated automation is the best choice, with the assembly cost being 0.5 (50%-99%) times that of the manual process.

Direct calculation

The relative costs of the different assembly processes may be compared by direct calculation. This method is fairly quick for a specific quantity, but is time-consuming if a number of alternatives are to be assessed. It is therefore beneficial to use a graphical method for the initial selection, and mathematical methods when exact and finalized information is needed. The data used for this calculation is set out below (Owen 1982):

1. Manual system:
   4 operators @ $8000/yr;
   single 8 hr shift;
   reject level of 5%;
   1 inspector @ $12 000/yr;
   allocated overhead $2000;
   cycle time 15 seconds.
2. Flexible automation:
   cost of system $80 000;
   depreciation 5 years straight line;
   cycle time 10 seconds.

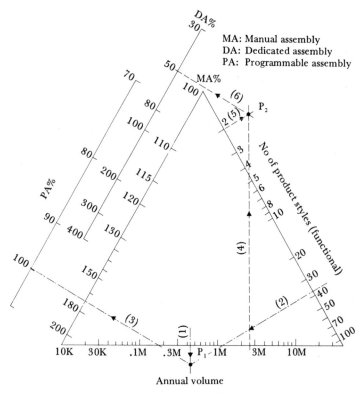

Figure 14.1 *This selection triangle is marked up for two test cases (1,2,3) and (4,5,6). The three parallel lines on the lefthand side of the triangle give the relative costs of the three alternative assembly methods.*
(Csakvary 1981)

3. Hard automation:
   cost of system $200 000;
   depreciation 5 years straight line;
   cycle time 1 second.
4. Product:
   annual product 480 000;
   labour content $0.10;
   material content $5.90;
   selling price $10.

For the initial comparison, it is assumed that inspection and allocated overheads are required and are the same for the three alternatives. Table 14.1 shows the various categories of costs, including their derived values. For a particular batch size, the payback period and the per unit cost are also computed.

Material	Manual	Flex. auto	Hard auto
System cost	$1,000	$80,000	$200,000
Depreciation	NA	$16,000	$40,000
Fixed costs	$14,000	$30,000	$54,000
Variable cost	$6.00	$5.90	$5.90
Selling price	$10.00	$10.00	$10.00
Breakeven	3,500	7,317	13,170

For a batch of 480,000 components

	Manual	Flex. auto	Hard auto
Production cost	$2,894,000	$2,862,000	$2,886,000
Cost reduction v manual	NA	$32,000	$8,000
Payback period	NA	2.50 yrs	25 yrs
Cost per assembly	$6.029	$5.9624	$6.0124

Table 14.1 *Categories of costs, including their derived values*

It can be seen from Table 14.1 that flexible automation offers the lowest cost per unit, with a payback period of two and a half years. What is of interest is to consider the other benefits that *could* be derived from the adoption of robots, and thence to determine the total expected savings.

Benefits and total expected savings

Additional benefits come from the increased consistency of quality synonymous with robotic assembly. There are six sectors where quality related savings can be expected:

1. A reduction in the amount of material and products that it will be necessary to scrap. Irrespective of the cause, there is always some irredeemable financial loss in terms of material, labour and energy costs, expended in processing items which are later scrapped because of quality related reasons.

2. A reduction in the cost penalty of lost revenue, incurred through assembling products that are later rejected and consequently cannot be sold.

3. A reduction in the warranty costs incurred by the manufacturer. If the assembly process can be defined in terms of quality assurance and consistency, the product's functional reliability can be determined. Hence the number of 'within warranty' failures can be predicted and the cost structure adjusted to compensate for this cost.

4. A reduction in the costs of inhouse inspection. If a production system yields products of varying quality, an external inspection facility is necessary so that the products can be tested and unsatisfactory ones can be rejected and/or returned for rectification. If a process guarantees products of a known and acceptable standard, inspection as a production task must be challenged. The incurred cost of inspection relates to the number of inspectors involved, as well as to the probability of rejected units being accepted by a client, together with the resulting cost of rectification and embarrassment to the manufacturer for that erroneous acceptance. Only if the financial risk exceeds the cost of inspection should inspection as a task external to the assembly process be considered.

5. An increase in acceptable quality output, with decreasing operating costs. Although an automated system might not be able to process an individual product any faster than a manual system, its output of acceptable products per shift is invariably much better. Increasing the output of a manual system involves

more people or multishift working, together with the additional supervisory costs. The output of a robotic system increases in proportion to the time it is operated, whilst its operating cost reduces inversely.
6. A more reactive assembly system which, through a predictable and consistent quality output, can cope with demand fluctuations for products by varying the usage time of the robot. This is better than the alternative of hiring, training, and firing people as the demand for products varies. Finally, because the output per unit time of a robotic system is predictable, there will be a reduction in both the number of orders lost through stock outs and the number of products and components needed to cover for scrap and/or erratic output.

The economic advantages of the proposed robot system can now be revised. There is a probability of 0.80 that a 20% reduction in the reject rate will occur. If this is not so, the reject rate will be reduced by only 5%. Therefore:

Expected savings = $((0.80 \times 0.20 \times 0.05) + (0.20 \times 0.05 \times 0.05))$
$\times$ 480 000 $\times$ \$5.90
= \$24 072

There is a probability of 0.50 that the extra output can be sold without recourse to extended stocking or reduction of price. The remainder will incur a penalty of \$1.00 each. Therefore:

Expected savings = reduction in rejects $\times$ 0.50 (\$4.10-\$1.00)
= \$6324

There is a probability of 0.4 that the inspection costs will be reduced by 75% and a probability of 0.6 that there will be a 50% reduction. Therefore:

Expected savings = $((0.40 \times 0.75) + (0.60 \times 0.50))$ $\times$ \$12 000
= \$7200

The anticipated savings from the above nonmanpower activities is \$37 596, which is 17½% greater than the savings from manpower alone. The anticipated total savings from this simplified analysis is \$69 596, which gives a payback period of 1.15 years.

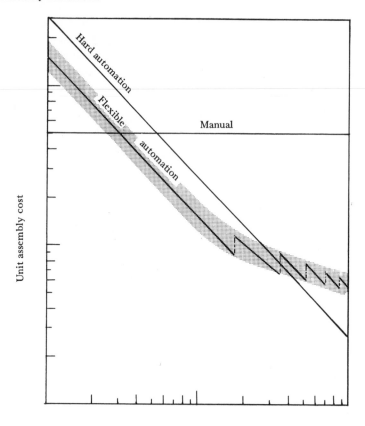

Annual production (millions)

Figure 14.2 *An econometric graph that shows the comparable cost (per unit assembled) for manual, flexible automation and hard automation techniques. The cost of manual assembly is usually assumed to be constant (piece work), whilst the flexible automation system's cost reduces constantly until the capacity of a single system is saturated. The cost line then exhibits a step function through the capacities of the adjacent systems. However, a smooth curve can be used to approximate the varying cost per item. Hard automation also shows a decreasing cost of production, but because the capacity of this third option is in excess of the other methods, the point at which it exhibits a change of shape because of saturation is outside the scope of this figure.*
(from *Flexible Assembly Systems*, Plenum Press, 1984)

Systems for a range of quantities

If alternative systems are compared on the basis of a single output quantity, no information is gained as to the sensitivity of the alternatives with respect to a range of quantities. A robotic assembly system is flexible in terms of what products it can process, within its parameters of geometry, power and function. Hence, the processing cost sensitivity to different quantities of different products is an important factor in the justification of an assembly system incorporating robots.

A graphical cost comparison of manual, flexible and hard automation alternative assembly systems is shown in Figure 14.2, which shows (for the particular data used) that the cost advantages of hard automation begin only when the annual demand is in excess of the capability of a single flexible alternative system.

For this mode of comparison, the manual cost is assumed to be fixed at the cost of processing a single unit, this being calculated on the basis of the operator's rate (or salary) and the time to complete that one assembly. It is acknowledged that for a specific quantity of products, the exact cost per assembly by a manual operator can be calculated.

Flexible and fixed aspects of automated assembly

When analysing the cost of using robots for assembly tasks, it is sometimes beneficial to make a distinction between the flexible and fixed equipment used within the assembly system. Some equipment (eg robots and general workstations) can be used for processing a wide variety of products, whereas other equipment (eg feeders, grippers and specific jigs) is dedicated to a particular product.

The expenditure of researching and installing an assembly system is a flexible expenditure, since the experience gained by the personnel involved can be used in other robot applications. The design cost of the system will have some fixed and flexible elements, depending upon how much of the knowledge is portable (Scott 1984).

SEGREGATED COSTS
The principle of segregated cost rates is given thus:

Unit cost = (cycle time) * ((cost per unit time of the flexible
part) + (cost per unit time of the fixed part))

It must be stressed that in practice the cost per unit time of the dedicated portions is straightforward to calculate, if the life of the fixed equipment is relatively short. However, the cost rate for the remaining equipment is far more difficult, since the cost of the robot should not be written off in less than at least five years. This is because its flexible nature allows it to be used for many more tasks than those for which it was originally purchased. Therefore, a robot has a useful life limited only by its technical obsolescence and its ongoing maintenance costs.

Although the mathematical model is very simple, its predictions nevertheless agree successfully with many more complex treatments, and it can answer such questions as:

1. Is it better to use a cheaper robot or a faster robot?
2. Which has the most effect on the unit cost?

The answers are readily obtained from this mathematical model. For instance, they show that halving the assembly time by doubling the robot speed has far more effect on unit costs than halving the cost of the robot. This is because a halved cycle time compensates for a costlier robot, and additionally halves the unit cost attributable to the fixed automation. However, halving the cost of the robot reduces only a portion of the cost of the flexible parts and none of the cost of the fixed parts of the assembly system. Also, since the assembly task is usually paced by the robot and not the material feeders, the system can cope with a faster robot.

Rogers (1978) agrees with the above statement, and says that even the most economic robot system must be justified by comparison with existing assembly methods. Also, that the economics of a robot system tend to be more sensitive to non-fixed costs (eg tooling and arm speed) than they do to fixed costs (eg those associated with arm hardware). An equation describing the assembly cost per component (in the product) can be summarized thus:

$$\frac{C}{n} = (t/2 + 2XT)\left[3W + \frac{W\left\{2R + N[C_c + nC_g + (ny + N_d)C_m]\right\}}{SQ}\right]$$

where C_c is the cost of the work carrier ($1000); C_g is the cost of the gripper per product component handled ($500); C_m is the cost of the manually loaded magazine ($500); N is the number of products (1); N_d is the number of design changes in the life of the assembly (0); Q is the equivalent cost of the operator in terms of capital equipment ($30 000); R is the cost of one robot ($40 000); S is the number of shifts (3); T is the system down time due to defective parts (0.01); n is the number of parts in the assembly (13); t is the mean time of assembly for one component (5 sec); and y is the number of styles for each product (1).

178

To illustrate the effects of fixed and nonfixed costs, the above bracketed values can be substituted into the equation. Further, if the cost of the robot is halved, the effect on the per unit price is insignificant. But if the robot 'speed' is doubled (which has the effect of halving the assembly time), the per unit cost is reduced by almost 40%, proving that the speed of operation is a more sensitive parameter for assembly cost. Consequently, the estimation of the cycle time becomes an important factor when evaluating alternative systems (see Chapter 4).

The structure of the segregated cost-rates principle demands that the user identifies which elements of the robot assembly system are fixed and which are flexible. This requirement gives the user the opportunity to review the make-up of the assembly system in the light of the batch sizes and products to be processed. Large batch size and small numbers of product types should be reflected by the higher usage of the fixed elements of the system. Conversely, a large flexible to fixed ratio of equipment will reflect a system that deals with small batch sizes and a large number of product types.

This rationale is proved by considering hard automation which processes large numbers of (usually) one product type, and contains a very high percentage of equipment that cannot be salvaged or reconfigured for another product. It is also substantiated by the ability of flexible manufacturing (and assembly) systems to process a single item economically. The cost of these systems is amortized over many thousands of items and is therefore negligible. The only costs that can be attributed are those of material, setup and operation.

Material cost, ignoring scrap and quality problems, will be the same by any method. The cost of setup varies with the flexibility of the system being used. If jigs and tools can be reconfigured with software, it is obviously quicker and cheaper to use them rather than having to perform a lot of hardware changes. If the machine can be quickly reconfigured and the tools changed automatically without causing downtime, the cost of operation will also be more attractive. Therefore a 'very' flexible system can process extremely small batch sizes economically.

CHAPTER 15

Economics of robots and grippers

The price of adopting a robot consists of three elements: the basic robot, its accessories, and the cost of installation. As shown in Figure 15.1, the breakdown of these costs varies with the primary application. For assembly tasks, the values are:

1. Basic robot being 40% of the total system cost.
2. The accessories being 35% of the total cost.
3. The installation being 25% of the total cost.

The hourly cost of using a robot has already been shown in Chapter 13 to be less than the *true* hourly cost of employing a person. Also the expected useful life of a robot within industry is long, compared with its use on any one assembly task or product range. Hence, a proposal for robotizing an assembly task should show the incurred costs for what they really are, so that the proposal can be assessed upon genuine facts, instead of popular myths.

Further, it is important to acknowledge that the value of the robot to the company will diminish very slowly, whilst the cost of the specialized gripper and auxiliary equipment will need to be recouped during the life of the initial project. Similarly, the cost of implementation must be placed totally against the particular project. For example, if the value of the robot is depreciated in a straight line to zero over five years, and the desired payback period and life of the project is two years, simplistically the cost to that project of the robot is 40% of the basic cost, plus the cost of the accessories and installation.

For the breakdown given above, the costs that have to be recouped are $(0.4 \times 40\%) + 35\% + 25\%$, which equals 76% of the total cost. Alternatively, this means that a \$100 000 project contains a \$24 000 cost element that does not have to be regained.

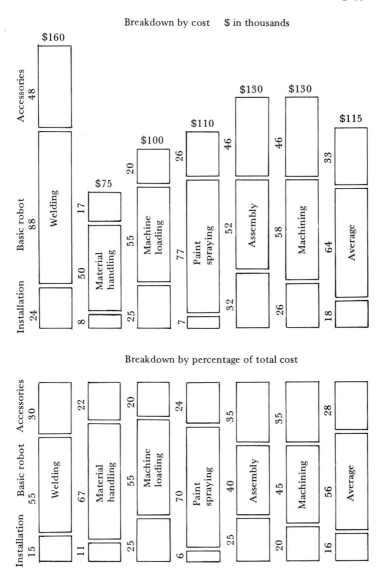

Breakdown by cost $ in thousands

Breakdown by percentage of total cost

Figure 15.1 *The total cost of any robot system is ill-defined and/or never computed. These listed values give dollar and percentage values for the three system elements of basic robot, accessories, and installation for typical robot systems.*
(from Economist Intelligence Report 135, *Chips in Industry*, 1982)

The workstation

A robotic assembly workstation can include one or several robots, each with a single or multiple gripper. The choice and cost of the robot-gripper combination to perform a given task in a given time frame can be assessed in a quantitative analysis as follows.

The analysis of 'n' grippers for a generalized workplace is considered in isolation so there is no constraint on that workstation's cycle time due to the activities of adjacent workstations.

For a given workstation, the total time taken in performing a number of actions is a function of the time taken for each action. Therefore, assuming steady state conditions:

$$t_c = \sum_{i=1}^{n} t_i$$

where t_c is the workstation cycle time, t_i is the time to perform the ith action and n is the number of actions.

From the above it can be seen that minimization of the nonproductive time within an action will result in a shorter cycle time. Minimization of nonproductive actions can be achieved through optimization of the workstation functions and/or the use of one or more grippers.

The workstation functions that could be optimized are associated with:

1. Component presentation, so the correct item is always in the proper place at the right time.
2. Automatic changeover of consumable material containers/dispensers that do not cause shutdown of the system.
3. Finely tuned interlocks that do not cause long delays before authorizing the next action.

The nonproductive actions associated with the gripper are those that, for the whole of their duration, do not perform value added work to the products being assembled at the workstation. These activities are:

1. Movement of the robotic arm with nothing in its gripper.
2. Movement of the robotic arm, while transferring subassemblies around the workstation and/or placing them onto the intercell transit system.

3. Movement of the robotic arm to and from the gripper storage station for gripper changeover.

It can be seen that the time occupied by the robotic elements of the workstation can be minimized by careful design of the workstation, several robots working in concert, a robot with multiple arms, or multifunction grippers.

WORKSTATION DESIGN
The design of the workstation is discussed in Chapter 8, whilst multiple robot cells are dealt with in Chapter 5, see Figure 5.11.

MULTIPLE ARM ROBOTS
There are two configurations of robots with multiple arms:

1. The YES-MAN, which is a true robot with two separate arms (discussed in Chapter 3 and illustrated in Figure 3.15).
2. The use of any of the 'nonrobots' fitted with multiple arms at fixed angular displacements (see Figure 5.12).

MULTIFUNCTIONAL GRIPPERS
Multifunctional grippers (discussed in Chapter 5) are rotated into position through software controls from the robot's controller, as and when they are required. In many cases, the individual grippers are easily released from the 'turret' and changed to suit different tasks, even though this is a manual task much like the changing of the tool rack on computer controlled machining centres.

Multi-arm, multirobot or multifunction grippers are only viable if the additional investment can be recouped through the derived savings. Simplistically, the case can be stated thus:

$$I \leqq P \sum_{i=1}^{n} (s_i \cdot c_i) - C_c - F - M$$

where C_c is the additional cost of control and programming per action, F is the additional direct and indirect cost factors per action, I is the additional investment at the workstation, M is the increased maintenance cost per action, P is the number of components produced within the payback period, c_i is the cost per unit time for the ith action, s_i is the time saving for the ith action, and n is the number of actions.

In addition to any quantitative rationalization of a generalized workstation, there are always a number of parameters that pertain to a particular workstation which must be investigated and resolved before the final decision regarding the complexity of the robotic workstation is made. These are:

1. The possibility of collision between the individual arms of the multirobot option. The solution is a function of the sophistication of the control system.
2. The size of the work envelope in which the multi-arms, and multifunction grippers will be operating. It must be remembered that access of one arm/gripper means that the others must be indexed out of the way, hence potential collision points must be identified and removed.
3. The processing of more than one type of component through the workstation, which may negate any cost benefit realized through using other than a simple robotic system.

The gripper design
Assembly is the most advanced, complex and sophisticated use of robots at the present time. However, to decrease the time and cost occupied by assembly tasks to a minimum, it is necessary to give the greatest attention to the design of the grippers.

Grippers are often designed to suit the geometry and complexity of a particular component. Wherever possible, the gripper should include some flexibility so it can be used for handling and manipulating *all* the components being processed by the robot. Unfortunately, such a universal gripper does not exist at the moment, though it is possible that it will be available in the foreseeable future.

The designs that come closest to satisfying the ideal of universality are those for artificial hands designed specifically for robots (Figure 5.7). These usually have three fingers – thought to give virtually the same capability as the human's five fingers. At present, these devices are more suited to experimental work than to the real world of industry and automated processes, such as assembly.

In Chapter 5, four gripper classifications are listed (see Figure 5.10), the most common and the most cheap being the flexible style. This type of gripper has built in mobility and adjustment which accommodate a wide range of different shaped and sized components. The special gripper is the exact opposite of this in that it is designed to perform a very specific task efficiently. Whilst the relative costs of manufacture depend entirely on the type of task to be performed, as a general case the special gripper is less expensive than the flexible.

A compromise between the special and flexible gripper is the multifunction unit, which contains a number of different gripper heads attached to a common base. Each of the gripper heads can be specially designed for a given purpose, or be flexible. This multiplicity of grippers and styles allows a robot to process a complex product without recourse to multiple workstations and/or changing single fixed grippers from a tool store. The cycle time is therefore the sum of the task times plus the time required to rotate the multifunctional gripper as needed to present the correct gripper to the work.

Martensen and Johansson (1980) discussed the economics of gripper design and determined, within the bounds of a generalized case, that the cost of assembly per unit increased more rapidly as the number of parts requiring exchange of grippers, than the number of parts themselves. Figure 15.2 substantiates this claim, where it can be seen that for a given number of parts the per unit cost increased in relation to the number of grippers used.

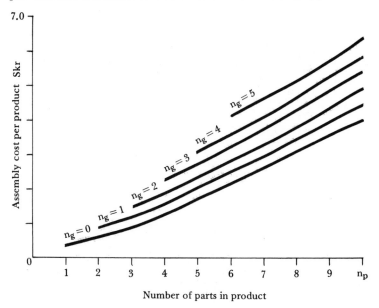

Figure 15.2 *Assembly cost per product varies as a function of the number of parts in the product(s) and the number of gripper exchanges required.* (from Martensen and Johansson 1980)

Figure 15.3 shows the relative costs incurred for the use of a number of separate grippers used to assemble a given product. The comparison is made on the assumption that the time taken to exchange the gripper is equal to the time taken to assemble a single part which, for the example shown, is eight seconds.

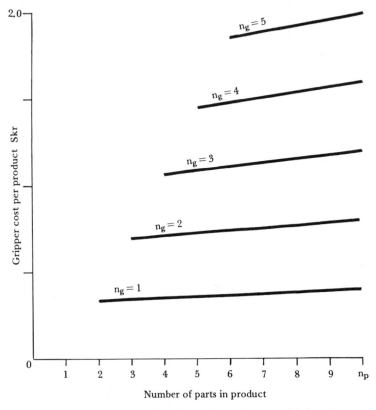

Figure 15.3 *The cost of grippers per product assembled varies as a function of the number of parts in the product(s) and the number of gripper exchanges required.*
(from Martensen and Johansson 1980)

If the number of gripper 'exchanges' is reduced from five to three per product, the cost saving per product varies from SKr 0.8 and SKr1, dependent upon the number of parts in the product. This means that the amount of effort spent upon designing a more complex multifunctional gripper can be quantified from the reduction in processing time derived from the new design. The equations used in the assessment are:

$$C = (T_a/T_i)I + C_e + C_h$$

$$I = C_r + C_g(n_g + 1) + C_p + n_p(C_s + C_f) + C_d$$

$$C_e = C_j \left\{ (n_p * n_f * T_i * L) + T_o + [T_t(1 - X) * L] \right\}$$

where C is the mean production cost to assemble a product; C_d is the cost of sensors, gauges, etc; C_e is the cost penalty of efficiency; C_f is the cost of feeding equipment per component; C_g is the cost per gripper; C_h is the cost of overhead (rents and operational costs); C_j is the cost of service personnel and operators; C_p is the cost of planning the system; C_r is the cost of the basic robot; C_s is the cost of storage for equipment; I is the investment; L is the number of service personnel; n_f is the percentage of faulty parts; n_g is the number of parts requiring an exchange of grippers; n_p is the number of parts in the assembled product; T_a is the mean assembly time per product; T_i is the idle time of the system caused by a faulty part; T_o is the time occupied by the system operator; T_t is the total available assembly system time; and X is the mean availability of the assembly system (percentage of total).

CHAPTER 16

The future

All forecasts are based upon a summary of known facts and logical deductions. The facts are:

1. There are a large number of robots available for use in assembly tasks.
2. The majority of assembly tasks in industry are not being performed robotically at present.
3. Robots are not being generally adopted by industry, let alone for use as assembly elements.

The cartesian and gantry robots, discussed in Chapter 3, offer the logical configuration for performing assembly tasks. Yet, of the 39 assembly robots commercially available in the USA and the UK, 26% are of the arm and elbow configuration and 41% are of the SCARA type (see Appendix). Furthermore, of the robots listed, 17 are Japanese, 15 are American, 2 are Swiss, 2 are Italian, 2 are British, and 1 is Swedish. It is acknowledged that Unimation assemble a number of their models at Telford, but their design was originally American.

So what are the reasons for the present day makeup of the UK's robot industry, and what can be deduced from it? A product, and after all that is what a robot is, sells on the basis of value for money, quality and the belief of the marketplace. Consider, for instance, the two gantry robots presently available for assembly tasks. Neither has the repeatability of other robots, even though structurally they are stiffer and simpler to control. Alternatively, of the two cartesian robots, the PRAGMA is 'popular', though expensive if only one arm is required. For more than one arm, its price becomes competitive with other robots, and with three or four arms it cannot be beaten for value or quality. The other cartesian robot, VS Remek's PAM, suffers from the 'don't buy British' attitude

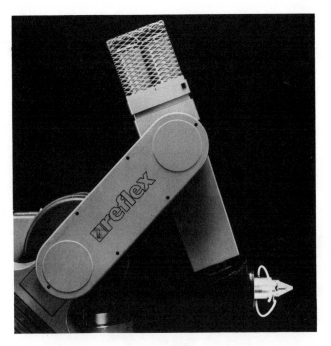

Figure 16.1 *American Robot Corporation's Merlin assembly robot, known as the Reflex in the UK. Its payload is 22kg and it has a maximum reach of 1016mm.*
(courtesy of Rediffusion Robot Systems Ltd)

which assumes that all UK manufactured goods are inferior to other nationalities. The consequence of this attitude has been the virtual demise of the indigenous robot and machine tool manufacturer.

The remaining 20% of the assembly robots available on the UK market is split evenly between those of cylindrical configuration and those of (at the moment) unusual configurations (eg the Microbro MR-1 and the new ASEA 1000). Hence, the market is dominated by robots with the configurations illustrated in Figures 16.1 and 16.2.

The present market for robots in general and for assembly robots in particular is relatively small, because manufacturing industries are reluctant to invest in new technology until the present recession ends. Many large corporations are investigating assembly robots, though the units tend to be purchased without any preconceived target in mind. Since the cost of the most sophisticated commercially available robot constitutes

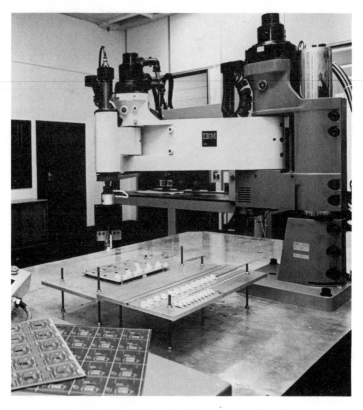

Figure 16.2 *The IBM 7540 SCARA configuration robot.*
One of the three marketed by IBM under primary licence from
Sankyo Seiki. These units are 'customized and refined' through
the inclusion of IBM control systems before being marketed.
(courtesy of IBM United Kingdom Ltd)

'petty cash' to these large corporations, the adoption of assembly robots by this small sector of the manufacturing industry should not be regarded as typical of the whole industry.

The smaller companies which make up the majority of the manufacturing industry are worried by the rapidly changing membership of the robot industry – they do not wish to purchase a robot and then discover that the vendor leaves the industry. Further, the large number of vendors, along with their hundred or so models, make it easier to postpone a purchase than to make the decision to buy.

The robot industry of the USA is far more established than the UK's. This is probably because two of the largest robot manufacturers are themselves American, and also there are a large number of smaller robot manufacturers. American industry is more aware of the advantages of using robots for assembly and, of course, the real cost of purchasing a robot is lower in the USA than in the UK. With a larger manufacturing base, supported by a large consumer population which enjoys a higher standard of living because of higher salaries, it is no wonder that the American adoption and use of assembly robots is forging ahead at an ever increasing rate.

There is a huge potential market for vendors if they can survive until market pressures consolidate the robot industry. It is perhaps with this objective in mind that new and innovative robots are being designed and manufactured for assembly-related tasks.

GEC's Marconi Research Laboratories have designed and are presently evaluating their advanced device for an assembly robot (GADFLY), shown in Figure 16.3. The triangulated structure was determined as the one most likely to produce the requirements for high speed of operation, adequate repeatability, and high structural stiffness.

Triangulated structures, in which each member is under either pure tensile or compressive stress, meet these needs admirably. In addition, the number and mass of the *moving* parts are reduced, again assisting in the primary targets of high speed and extreme rigidity. The position and attitude of the gripper are controlled by the movements of the threaded rods through the joints at the top of the robot. Adjustment of the rods can be performed simultaneously, and small motions of the rods yield a rapid change in gripper attitude. Hence, operational speed is gained through small simultaneous motions.

ASEA's new robot (Figure 16.4) was launched at the 14th ISIR conference at Gothenburg, Sweden in October 1984. This six axis robot has a payload of 3kg and is claimed to be the fastest in the world, with an arm velocity of almost 12m/sec. Its unique feature is that the vertical arm is fitted as a pendulum, with its pivot point at the centre of gravity of the moving elements in the robot. This arrangement is claimed to allow acceleration rates 50% faster than those possible with a 'conventional' robot, with a resulting reduction in task times.

If one of the future attributes of assembly robots is that of speed of operation, the Microbro's MR-03 (Figure 16.5) should be considered. This recently launched unit is double

Figure 16.3 *Marconi Research Laboratory's GADFLY robot.*
The gripper is moved in attitude, orientation and plane
through relative motion of the six threaded rods.
© *The General Electric Company plc.*
(courtesy of Marconi Research Laboratory)

headed, is designed for small part assembly, and can be mounted either on the worktable or on an overhead gantry.

With some idea of what is happening at the moment in the industrial marketplace for assembly robots, the future prospects for these devices can be viewed in both the short and the long term. The final sections of this chapter can also be interpreted as the tactical and strategic plans of the robotics and manufacturing industries of the western world.

The short-term outlook

The short term (until mid 1986) will see a rationalization of the number of vendors in the robot marketplace and an increase in the number of products being designed for automatic assembly.

Figure 16.4 *ASEA's IRb 1000 robot could be called a seven axis gantry robot. It consists, however, of two independent modules, one being a six axis robot module and the other the single axis powered horizontal FAST Track 1000, which can be up to 11.25m in length. An alternative configuration is for the robot module to be mounted on a pedestal or hung from a gantry.*

The robot module holds the vertical arm in a yoke that permits rolling and pitching of the arm relative to the horizontal track. The third axis of motion is that of arm extension, whilst the wrist has its own independent three motion axes of roll, pitch and yaw.

The pendulum action of the robot module allows a repeatability better than 0.10mm, whilst that for motion along the gantry is 0.05mm. In addition, very high velocities and accelerations are claimed that will result in a significant reduction in assembly task times.
(courtesy of ASEA Robotics, Västerås)

The large number of robot vendors and the relative size of the market have meant that nobody is making any money from selling robots at present. Consequently, the number of vendors of robots, as well as the number of robots available, will be dramatically reduced. In fact, this is well overdue. When market forces prevail and the rationalization of both vendors and products is complete, the survivors will have a ready market. The size of the market, together with the increased market share, will enable vendors to give a reliable and high quality service to the manufacturing industry.

One short-term effect of the market uncertainty will be that vendors will start to charge for quotations. The reasons are twofold:

1. The marketplace has become accustomed to getting multiple quotations for each idea that is advocated as being potentially suitable for robotic assembly.
2. The take-up rate of these quotations is abysmal, being typically about 1%-5%. Also the time scale between quotation and placing of any order is between 6 and 18 months, by which time another quotation is usually needed.

The cost of robots will duplicate the trend shown in personal computers by rapidly dropping in terms of a given unit of capability. Looking solely at assembly robots, one technical change will take the form of a wide sensory capability supplied as an integral element, both in the robot's control system and in the form of grippers incorporating a high level of intelligence. In this way, robots will be able to cope economically with a randomized dynamic workplace, with the design of jigs and other retention devices being made as simple and universal as possible.

Another technical change will be an increase in the trend towards offline programming, especially through the use of personal computers. The development of artificial intelligence and expert systems will enable tasks to be specified objectively, without the need for very detailed work by the programmer/supervisor. This economically forced separation between the robot and its programmer will be encouraged through the wider use of CAD/CAM systems and the increased intelligence of the assembly robots.

Products will be designed for automated assembly irrespective of the batch size, and there will be an increased usage of honeycomb materials, plastics and high strength to weight ratio materials. The short-term future will see the demise of the discrete fastener, its place being taken by adhesives, moulded snap fasteners and ultrasonic or laser welding. In other words, movement will be towards assembly techniques easily automated, which will reduce, in addition, the numbers of components with a subsequent reduction in prime cost.

Figure 16.5 *The Microbro MR-03 electric drive, double headed robot has a payload of 1kg and a maximum reach of 210mm.*
(courtesy of Concentric Production Research Ltd)

The long-term outlook

The longer term (mid 1986 to 1995) outlook for assembly robots has two scenarios: one optimistic, the other pessimistic. The pessimistic scenario shows that the decline of the manufacturing industry will continue and that the western world's economy will collapse, since it can neither compete with other manufacturers nor be self-sufficient. In many ways, this scenario is the self-fulfilling prophecy presented in particular by the UK manufacturing industry, which cannot come to terms with the need for a reduced manual workforce, reduced hard automation and increased flexible automation, so that the fluid market pressures can be rapidly responded to in both demand and price. Obviously, within this scenario, there is no future for assembly robots.

The optimistic scenario shows the inevitable migration of people from the manufacturing industries to those of service, information and leisure. A manufacturing industry that uses the western world's information technology will be able to compete with the low cost manual intensive industries from the Third World. An analogy can be seen in the western agricultural industry, which has moved from being a manually intensive process to a high technology industry, employing a very small percentage of the population. The present day agricultural industry is much more productive than its historical self, or the present day agricultural practices of the second and third worlds. It freed a large percentage of the population to do other jobs, yet more than satisfied the market demand for agricultural products. In the same way, it is anticipated (hoped) that the transition from a manufacturing-based economy to an information-based economy will be equally successful.

This scenario sees the present development of the flexible manufacturing and assembly systems (FMASs) into computer automated factories (CAFs). This computer integrated manufacturing (CIM) will enable orders to be sent directly from a client to the CAF by computer, which will then automatically schedule and process the order. The CIM system will include a totally automated CAD/CAM system that will identify or design the tooling, configure the jigs and call up or write the programs for downloading to the computer controlled conversion sector of the CAF. When the order is complete, it will be shipped directly to the client, whose computer will credit the supplier against the computerized invoice.

Obviously, machines within the conversion sector will have to include a large amount of intelligence, so that they can be operated 24 hours a day without human supervision. Computerized monitoring and automatic tool-changing already exist, together with automatic material identification and transportation. Material technology will be developed so that processes like net shape forming become commonplace, resulting in a number of material removal processes being minimized. Hence, the remaining processes are those of assembly and inspection, which have already been shown to be viable tasks for robots of a particular technology. This means that the demand for assembly robots will increase, whilst that for other robots will decrease.

The hard automation alternative to assembly robots requires a large product demand to ensure viability. However, the fickleness of today's market will be increased in the future by the need to satisfy the almost individual demand patterns of the population.

Although the future will most likely turn out to be somewhere between the two scenarios described above, the need for robots to perform assembly tasks is assured. Similarly, the need to increase their sophistication and capability will occupy people for many years to come. Assembly is a task that is best performed by robots, leaving humans to perform tasks that give personal satisfaction and to make best use of their innate intelligence and skills.

References and bibliography

ACARD (1979) *Joining and Assembly: The Impact of Robots and Automation* Her Majesty's Stationery Office, London.

Aleksander, I.; Schlesinger, R.J. (1983) *The International Robotics Yearbook* Kogan Page, London.

Atkinson, B.; Heywood, P. (1982) *The Challenge of Vision* British Robotic Systems Ltd, London.

Ayres, R.U.; Miller, S.M. (1983) *Robotics: Applications and Social Implications* Ballinger, Cambridge, Massachusetts.

Boothroyd, G.; Dewhurst, P. (1983) *Design for Assembly: A Designer's Handbook* University of Massachusetts.

Coiffet, P. (series editor and consultant); Aleksander, I. (English language series consultant) (1983-85) *Robot Technology* (an eight-volume series plus index and bibliography) Kogan Page, London.

Csakvary, T. (1981) *Product Selection Procedure for Programmable Automatic Assembly Technique* Proc. 2nd. Int. Conf. on Assembly Automation, Brighton, UK, pp 201-210.

Kehoe, E.J. (1984) Robotic workcell combines ultrasonic insertion and assembly operation, *Robotics Today*, pp 93-95, April.

Kondoleon, A.S. (1976) Application of technology-economic model of assembly techniques and programmable assembly machine configurations. Master's thesis, Massachusetts Institute of Technology.

Lundstrom, G.; Glemme, B.; Rooks, B.W. (1977) *Industrial Robots-Gripper Review.* IFS (Publications) Ltd, Kempston.

Martensen, N.; Johansson, C. (1980) *Subassembly of a Gearshaft by Industrial Robot*, 10th Int. Symp. and 5th Int. Conf. on Industrial Robots, Milan, Italy, pp 523-533.

Naruki, K. (1981) *How to Promote Automation in Japanese Industries* Proc. 2nd Int. Conf. on Assembly Automation, Brighton, UK, pp 143-151.

Nevins, J.L.; Whitney, D.E. (1979) In *Computer Vision and Sensor Based Robots* (*eds* G.G.Todd and L.Rossell) pp 275-321, Plenum Press, New York.

Open University (UK) (1984) *Three-volume Textbook for the Robotics Module of a Master Degree Course in Manufacturing Systems.*

Owen, A.E. (1982) *Automated Assembly can Equate with Short Payback Periods.* Proc. 3rd Int. Conf. on Assembly Automation, Stuttgart, West Germany, pp 353-362.

Owen, A.E. (1982) *Chips in Industry.* Economist Intelligence Unit, London.

Owen, A.E. (1983) Designing for automated assembly, *Engineering*, June.

Owen, A.E. (1984) *Flexible Assembly Systems* Plenum Press, New York.

Owen, A.E. (1984) Guidelines to the selection of robots, *The Industrial Robot* 1 (11) pp 25-28.

PERA (1983) Robots: a further survey of robots and their current applications in industry, Report 377, Melton Mowbray, UK.

Railbert, M.H.; Tanner, J.E. (1982) Design and implementation of a VLSI tactile sensing computer, *The International Journal of Robotics Research* Vol 1 No 3 pp 3-18.

Redford, A.; Swift, K. (1980) Salford University Industrial Centre Report, August.

Rogers, P.F. (1978) A time and motion method for industrial robots, *The Industrial Robot* 5 (4) pp 187-192.

Scott, P.B. (1984) *The Robotics Revolution* Basil Blackwell, Oxford.

Sutton, G.P. (1980) The many faces of productivity, *Manufacturing Engineering* 85 (6) pp 58-64.

UK Department of Industry (1977) *Commitment for Terotechnology Life Cycle Costing in the Management of Assets: A Practical Guide.*

University of Nottingham (1982) Proceedings of one-day seminar on robot safety.

Appendix

Assembly Robots Available in the USA and the UK (correct at time of printing)

Vendor	Robot name	Basic cost	Configuration	Repeatability	Programming style	Country of origin
Adept Technology Inc Mountain View CA 94043 (415) 965-0557	Adept One	$40,000	SCARA	0.05mm	Offline	USA
Meta Machines Ltd Oxford OX14 1DY 0235 22155	Adept One	$49,650 (priced in dollars)				
American Robot Corp Clinton PA 15026 (412) 262-2085	Merlin	$65,000	Arm and elbow	0.025mm	Online and offline	USA
Rediffusion Robot System Crawley RH10 2PY 0293 543255	Reflex	£42,000				
ASEA Inc Troy MI 48084 (313) 528-3630	IRb 1000	N/A	Novel gantry and pendulum	0.10mm	Online (offline in future)	Sweden
ASEA Ltd Milton Keynes MK13 9HA 0908 319666	IRb 1000	£28,000				

Assembly with Robots

Vendor	Robot name	Basic cost	Configuration	Repeatability	Programming style	Country of origin
Automatix Inc Billerica MA 01821 (617) 667-7900	AID 600	$70,000	Cartesian	0.07mm	Online and offline	USA
Control Automation Inc Princeton NJ 08540 (609) 799-6026	CAR-2000 Mini-Sembler	$40,000	Cartesian	0.02mm	Online and offline	USA
Fleximation Systems Corp Burlington MA 01803 (617) 229-6670	Toshiba Tosman TSR 701H	N/A	SCARA	0.05mm	Online and offline	Japan
Evershed Robotics Ltd Chertsey KT16 8LJ 09328 61181	Toshiba Tosman TSR 701H	c £20,000				
GCA Corp Industrial Naperville IL 60566 (312) 369-2110	DKP 200H DKP 200V	$24,000 $30,000	SCARA			
Dainichi-Sykes Robotics Preston PR5 8AE 0772 322444	Daros PT200H PT200V	N/A	SCARA	0.05mm	Online	Japan
General Electric Company Bridgeport CT 06602 (203) 382-2876	Model A12	N/A	Cartesian			
Fairey Automation Ltd Swindon SN2 6HB 0793 615111	DEA Pragma A 3000	Varies on number of arms and grippers	Cartesian	0.025mm	Online and offline	Italy

Vendor	Robot name	Basic cost	Configuration	Repeatability	Programming style	Country of origin
GMF Robotics Corp *Troy MI 48098* *(313) 641-4242*	*Fanuc* *A00* *A0* *A1*	*N/A* *N/A* *N/A*	Cylindrical	0.05mm	Online	Japan
600 Fanuc Robotics Ltd Colchester CO2 8LE 0206 868848	*Fanuc* A00 A0 A1	£13,200 £18,400 £23,300				
Hirata Corp of America *Indianapolis IND 46240* *(317) 846-8859*	*Armbase*	*$15,300* *to* *$21,300*	SCARA	0.05mm	Online and offline	Japan
Airstead Industrial Brighton BN1 4GH 0273 606918	AR-H250 AR-H300 AR-450	£17,500 £19,500 £21,000				
IBM Advanced Manufacturing *Systems* *Boca Raton FL 33444* *(305) 998 7066*	*7535* *7540* *7545* *7565*	*N/A* *N/A* *N/A* *N/A*	SCARA SCARA SCARA Cartesian	0.05mm 0.05mm 0.05mm 0.10mm	Online and offline	Japan Japan Japan USA
IBM UK Ltd Croydon CR9 6HS 01-484 0621	7535 7540 7545 7565	£21,000 £26,000 £30,000 £98,000				

Vendor	Robot name	Basic cost	Configuration	Repeatability	Programming style	Country of origin
Intelledex Inc Corvallis OR 97333 (503) 758-4700	605S 605T 705T	$48,000 $48,000 $60,000	Approximate arm and elbow	0.025mm	Online and offline	USA
Zehntel Ltd Milton Keynes MK9 2NJ 0908 606655	Intelledex 605S 605T 705T	Between £40,000 and £60,000				
Machine Intelligence Corp Sunnyvale CA 94086 (408) 737-7960	Hitachi A3020	N/A	SCARA	0.05mm	Online	Japan
GEC Robot Systems Ltd Rugby CV21 1BU 0788 2144	Hitachi A3020	N/A				
Micro Robotics Systems Lowell MA 01854 (617) 937-1970	MR-01 MR-03	$40,000 $50,000	Special Cylindrical	0.02mm	Online	Switzerland
Concentric Production Sutton Coldfield B72 1RD 021-355 1266	Microbro MR-01 MR-03	£25,000 £30,500				

Vendor	Robot name	Basic cost	Configuration	Repeatability	Programming style	Country of origin
Pental of America Ltd Torrance CA 90503 (213) 320-3831	HR 26 HR 63	$20,000 $21,500	*(see below for details of HR 26 and HR 63)*			
Sale Tilney Technology Weybridge KT15 2RH 0932 48311	Pental HR 26 HR 63 Type 3 Type 4	£18,000 £21,300 £18,000 £21,300	SCARA SCARA SCARA SCARA	0.05mm 0.05mm 0.05mm 0.05mm	Online and offline	Japan
Unimation Inc Shelter Rock CT 06810 (203) 744-1800	Puma 250 260 550* 560* 760* Unimate Series 100	N/A $41,000 $41,000 $47,000 $60,000 N/A	Arm and elbow Arm and elbow Arm and elbow Arm and elbow Arm and elbow	0.05mm 0.05mm 0.10mm 0.10mm 0.20mm	Online and offline	USA and UK
Unimation (Europe) Ltd Telford TF3 3AX 0952 618931	Puma 250 260 550* 560* 760* Unimate Series 100	£21,750 £23,750 £19,850 £21,850 £47,000 N/A	SCARA	0.025mm		USA

205

Vendor	Robot name	Basic cost	Configuration	Repeatability	Programming style	Country of origin
United States Robots *King of Prussia PA 19406* *(215) 768-9210*	*Maker 100*	*$40,000*	Arm and elbow	0.05mm	Online and offline	USA
British Olivetti Ltd Milton Keynes 0908 311555	Sigma Series 3	N/A	Gantry	0.10mm	Online	Italy
John Brown Automation Coventry CV3 2NY 0203 450810	Wickman W500	£15,500	Cylindrical	0.05mm	Online	UK
VS Remek Ltd Milton Keynes MK2 3HY 0908 649494	PAM 2	£15,000	Cartesian	0.05mm	Online	UK

* These robots are included for completeness only, to indicate the total Puma range.
** Some manufacture and assembly of these robots is done in the UK. The amount of UK centered fundamental design work is unknown to the author.

Please note that no inference is intended to be drawn from either inclusion or exclusion of robots or vendors in this appendix. The dynamics of the robotic market mean that both products and vendors are entering and leaving the scene at a rapid rate. Consequently, any list only represents a 'snapshot' in time of the members of the assembly robots' fraternity.

Please note also that the italicized vendors are American. The UK vendors are shown beneath in normal type.

Index